Annals of Mathematics Studies

Number 64

THE EQUIDISTRIBUTION THEORY OF HOLOMORPHIC CURVES

BY

HUNG-HSI WU

PRINCETON UNIVERSITY PRESS

AND THE

UNIVERSITY OF TOKYO PRESS

PRINCETON, NEW JERSEY

1970

L.C. Card: 78-100997
S.B.N.: 691-08073-9
A.M.S. 1968: 3061

Published in Japan exclusively by the
University of Tokyo Press;
in other parts of the world by
Princeton University Press

Printed in the United States of America

PREFACE

These are the notes for a course on the Ahlfors-Weyl theory of holomorphic curves which I gave at Berkeley in the Winter quarter of 1969. This is a subject of great beauty, but its study has been neglected in recent years. In part, this could be due to the difficulty of Ahlfors' original paper [1]; a subsequent poetic rendition of Ahlfors' work by Hermann Weyl [7] does not seem to be any easier. The modest goal I set for myself is to give an account of this theory which may make it more accessible to the mathematical public.

My audience consisted of differential geometers, so these notes are uncompromisingly differential geometric throughout. I should like to think that differential geometry is the proper framework for the understanding of this subject so that I need make no apology for being partial to this point of view. On the other hand, I must add a word of explanation for the length of these notes which some readers would undoubtedly find excessive. The reason is that great care has been taken to prove all analytic assertions that are plausible but nonobvious, e.g. that certain constants in an inequality are independent of the parameters or that certain functions defined by improper integrals are continuous. Although the experts might think otherwise, I cannot help feeling that given a subject as intricate as this one, it is best to check through all the details rather than to let the correctness of the final conclusions rest on wishful thinking.

I assume that the reader knows a little bit about differential geometry, complex manifolds and complex functions of one variable, but not much of any of these is actually needed. It should be pointed out that Chapter II is essentially independent of the rest and gives a complete exposition of the Nevanlinna theory of meromorphic functions defined on open Riemann surfaces. The pre-requisites for this chapter consist merely of the most rudimentary knowledge of classical function theory and the differential geometry of surfaces. Chapter I is a disjointed collection of facts needed for the later chapters. If the reader survives this chapter, he should encounter no difficulty in reading the remainder of these notes.

It remains for me to thank Ruth Suzuki for an impeccable job of typing.

H. W.

INTRODUCTION

By a holomorphic curve, we mean a holomorphic mapping $x: V \to P_n\mathbb{C}$, where V is an open Riemann surface and $P_n\mathbb{C}$ is the n-dimensional complex projective space. The central problem of the equidistribution theory of holomorphic curves, crudely stated, is the following: given m hyperplanes of $P_n\mathbb{C}$ in general position, does $x(V)$ intersect any one of them? The motivation for this question comes from two different sources. The first is algebraic geometric: Suppose instead of an open V, we let x map a compact Riemann surface M holomorphically into $P_n\mathbb{C}$, then $x(M)$ is an algebraic curve and it is a matter of pure algebra to check that $x(M)$ must intersect every hyperplane of $P_n\mathbb{C}$. Thus the replacement of a compact M by an open V has the effect of transferring the whole problem from algebra to the domain of analysis and geometry. The second motivation comes from the case $n = 1$. $P_1\mathbb{C}$ is of course just the Riemann sphere and the above question becomes: given m distinct points of the Riemann sphere, can $x(V)$ omit them all? The celebrated theorem of Emile Picard says that if $V = \mathbb{C}$, then $x(\mathbb{C})$ cannot omit more than two points or else it is a constant map. It seems therefore entirely natural to seek an n-dimensional generalization of this remarkable result.

Yet the Picard theorem, like the above question, must be considered relatively crude in that it is only concerned with the extreme behavior of a point being omitted by the image of x. Equidistribution theory, on the other hand, is much more refined and delicate as it seeks to yield information on how often each individual point is covered or how often each individual hyperplane is intersected by x. Let us be more explicit: we will first explain this for the case of a meromorphic function (i.e. a holomorphic $x: V \to P_1\mathbb{C}$) and then go on to do the same for holomorphic curves in general.

On the outset, it is quite obvious that some restrictions must be placed on V before meaningful statements can be made. It has been determined elsewhere ([8], Part B) that the most suitable condition to impose on V is that it carries a <u>harmonic exhaustion</u>, i.e. that there exists a C^∞ function $\tau: V \to [0,\infty)$ such that (i) τ is proper (i.e. τ^{-1}(compact set) = compact set) and (ii) τ is a harmonic function outside some compact set of V. Then $V[r] = \{p: p \in V, \tau(p) \leq r\}$ is a compact subset of V for each r. (Example: If $V = \mathbb{C}$, then such a harmonic exhaustion on $\mathbb{C}$ can be chosen to be a C^∞ function which equals $\log r$ outside the disk of radius three, say. Then for all large values of r, $\mathbb{C}[r]$ is just the disc of radius e^r. In the same way, such a harmonic exhaustion can be chosen on any V obtained from a compact Riemann surface by deleting a finite number of points. Note that what we have defined should properly be called an <u>infinite</u> harmonic exhaustion; see Definition 2.4 of Chapter II, §6. By a theorem of Nakai (Proc. Jap. Acad. 1962, 624-9), the Riemann surfaces carrying an infinite harmonic exhaustion are exactly the parabolic ones.) One of the basic quantities in this theory is the <u>counting function</u> $N(r,a)$, defined as follows.

Let $n(r,a)$ = the number of points in $x^{-1}(a) \cap V[r]$, where $a \in P_1\mathbb{C}$. If we fix an r_0 so that τ is harmonic outside $V[r_0]$, then by definition

$$N(r,a) = \int_{r_0}^{r} n(t,a)dt.$$

For the definition of the second basic function, we note that $P_1\mathbb{C}$ carries the classical spherical metric, which is a conformal (hermitian) metric of constant Gaussian curvature. If we denote its volume form by ω, and normalize it so that $\int_{P_1\mathbb{C}} \omega = 1$, then the order function $T(r)$ is by definition

$$T(r) = \int_{r_0}^{r} dt \int_{V[t]} x^* \omega.$$

As $r \to \infty$, $N(r,a)$ obviously measures how often the point a is covered by the points of V. On the other hand, $T(r)$ measures the average coverage of the points of $P_1\mathbb{C}$ by V; in mathematical terms, we have the following theorem:

$$(0.1) \qquad T(r) = \int_{P_1\mathbb{C}} N(r,a)\omega(a).$$

In other words, T is the arithmetic mean of N. With this in mind, we introduce the defect function on $P_1\mathbb{C}$:

$$\delta^*(a) = \liminf_{r\to\infty}\left(1 - \frac{N(r,a)}{T(r)}\right).$$

It will follow from a later result that $0 \leq \delta^* \leq 1$. If $a \in P_1\mathbb{C} - x(V)$, then of course $N(r,a) = 0$ for all r, so that $\delta^*(a) = 1$. According to (0.1), the other extreme of $\delta^*(a) = 0$ is to be interpreted as that the point a is covered by the points of V as often as the "average" point of $P_1\mathbb{C}$. The main theorem of the whole theory can now be stated: suppose V has finite Euler characteristic, then given any m distinct points $a_1,\ldots,a_m$ of $P_1\mathbb{C}$,

$$(0.2) \qquad \Sigma_i\, \delta^*(a_i) \leq 2 + \chi,$$

where χ is a finite constant, which vanishes if $V = \mathbb{C}$ or $\mathbb{C} - \{0\}$. A first consequence is Picard's theorem. A second consequence is that $\delta > 0$ only on a countable subset of $P_1\mathbb{C}$. Thus except on a countable subset of $P_1\mathbb{C}$, the values of V are equally distributed over $P_1\mathbb{C}$, and hence the name "equidistribution".

There is also a beautiful by-product of the theory: the number "two" in Picard's theorem turns out to be accountable for by the Euler characteristic of the Riemann sphere. To put this in the proper context, let us go into the mechanics of the proof of (0.2) in some detail. First note that we may regard $P_1\mathbb{C}$ as the quotient space of $S^3 \subseteq \mathbb{C}^2$, S^3 being the unit sphere, by defining $(z_0,z_1) \in S^3$ to be equivalent to $(w_0,w_1) \in S^3$ if and only if $(z_0,z_1) = (e^{i\theta}w_0,e^{i\theta}w_1)$ for some θ. The equivalence class of (z_0,z_1) we denote by $[z_0,z_1]$. If $a = [z_0,z_1]$, then $a^\perp = [-\bar{z}_1,\bar{z}_0]$ is the point of $P_1\mathbb{C}$ orthogonal to a, orthogonal in the sense that the inner product $\langle a,a^\perp\rangle$ in terms of representatives is zero. One checks that $u_a(p) = \log \dfrac{1}{|\langle a^\perp,p\rangle|}$ is a well-defined C^∞ function in $P_1\mathbb{C} - \{a\}$ and that

$$\frac{1}{2}\, dd^c u_a = \omega \quad \text{in} \quad P_1\mathbb{C} - \{a\}, \tag{0.3}$$

where $d^c = \sqrt{-1}(d'' - d')$. An application of Stokes' theorem to (0.3) yields:

$$n(r,a) + \frac{1}{2\pi}\int_{\partial V[r]} d^c x^* u_a = \int_{V[r]} x^*\omega. \tag{0.4}$$

We shall always assume that $r \geq r_0$, so that τ is harmonic on $\partial V[r]$. In that case,

$$\int_{\partial V[r]} d^c x^* u_a = \frac{d}{dr}\int_{\partial V[r]} x^* u_a {*d\tau}$$

where $*d\tau$ denotes the conjugate differential of $d\tau$. Using this fact, we integrate (0.4) with respect to r to arrive at the first main theorem (FMT):

$$N(r,a) + \frac{1}{2\pi}\int_{\partial V[t]} x^* u_a {*d\tau}\Big|_{r_0}^{r} = T(r), \tag{0.5}$$

where we have used the notation $h(r)\Big|_{r_0}^{r} = h(r) - h(r_0)$. It can be shown that $\int_{\partial V[t]} x^* u_a {*d\tau}$ is positive and is continuous in a. This together with (0.5) leads to an important inequality:

(0.6) $N(r,a) < T(r) + c$, c independent of r and a.

From (0.6), one deduces that the defect function satisfies $0 \leq \delta^* \leq 1$, a fact mentioned above.

(0.5) relates N to T. We need to relate T also to the topological data of V and this is the content of the second main theorem (SMT). We need some definitions before we can state it. The holomorphic mapping $x: V \to P_1\mathbb{C}$ is ramified at a finite number of points $\{p_1,\ldots,p_m\}$ inside the compact set $V[t]$. If x is locally m_j to 1 in a neighborhood of p_j, we call (m_j-1) the stationary index of x at m_j. Let

$$n_1(t) = \Sigma_{j=1}^{m} (m_j - 1),$$

and we define $N_1(r) = \int_{r_o}^{r} n_1(t)dt$. Next, if we denote by $\chi(t)$ the Euler characteristic of $V[t]$, then we define $E(r) = \int_{r_o}^{r} \chi(t)dt$. Finally, on $V - V[r_o]$, we introduce the nonnegative function h by:

$$x^*\omega = h\, d\tau \wedge * d\tau.$$

Then the SMT states that

$$(0.7) \qquad E(r) + N_1(r) - 2T(r) = \frac{1}{4\pi} \int_{\partial V[t]} (\log h) * d\tau \Big|_{r_o}^{r} .$$

The proof of (0.7) makes use of the Gauss-Bonnet theorem. The coefficient 2 in front of $T(r)$ is the Euler characteristic of the Riemann sphere and, eventually, this 2 makes its way into (0.2).

The situation is now quite clear: (0.5) and (0.7) contain all of the information we need to derive the defect relations with the exception of the two line integrals $\int_{\partial V[t]} x^* u_a * d\tau$ and $\int_{\partial V[t]} (\log h) * d\tau$. What we should do now is to obtain an upper bound for their sum in terms of $T(r)$, which is roughly of the form

$$(0.8) \qquad \int_{\partial V[r]} x^* u_a * d\tau + \int_{\partial V[r]} (\log h) * d\tau < C \log T(r).$$

A substitution into the left side the values of the line integrals as given by (0.5) and (0.7) would then yield immediately (0.2). The basic idea to obtain the inequality (0.8) is to integrate the inequality (0.6) over $P_1\mathbb{C}$ with respect to an appropriately chosen function ρ on $P_1\mathbb{C}$. It is a virtue of this geometric approach that the choice of ρ turns out to be entirely natural: $\log \rho$ should be roughly u_a.

The above theory was created by R. Nevanlinna in 1925. In the words of Hermann Weyl, this contribution of Nevanlinna constitutes one of the great mathematical achievements of the century. The simplicity of the modern proof of (0.2) we owe largely to Ahlfors; the beautiful idea of invoking the Gauss-Bonnet theorem in Nevanlinna theory is also due to Ahlfors. The exposition of Nevanlinna theory given in Chapter II follows closely that of [8], which in turn rests on the efforts of Ahlfors and Chern. It should be pointed out that instead of mapping into the Riemann sphere, one could equally well have chosen any compact Riemann surface as the image space (see [8]), but there is no doubt that the Riemann sphere occupies the central position of this theory. In addition to this, there are two reasons why I have exposed this special case with such deliberate care. The first is that it gives us some insight into the structure of the general equidistribution theory of holomorphic curves, which would otherwise have been lost sight of in a maze of technical details. More importantly, I feel that there are a few obvious open problems in this direct that are worth looking into. For one thing, the Nevanlinna theory on the parabolic surfaces with infinite Euler characteristic has hardly gotten started. Also, the relationship between the defect function and the choice of the harmonic exhaustion should be clarified. Finally, of the many remarkable results established for meromorphic functions on $\mathbb{C}$ in the last decade (see [4] and a survey article in Bulletin Amer. Math. Soc. 1967, 275-293 by W. Fuchs), it seems that most would survive on those V obtained from a compact Riemann surface by deleting a finite number of points. I hope that this exposition of Nevanlinna theory will stimulate someone into performing the task of transplanting these theorems from $\mathbb{C}$ to this class of Riemann surfaces.

The equidistribution theory of holomorphic curves was first attempted by H. and J. Weyl in 1938, and was brought to essential completion by Ahlfors in 1941. A short history of the subject has been written by Hermann Weyl in his usual inimitable style in the preface to [7] and, despairing of bettering Weyl's lyricism, I will simply refer the reader to that monograph. The exposition of this theory in these notes follows the general guideline of [8]. Contrary to the practice of Weyl and Ahlfors, we deal directly with $P_n\mathbb{C}$ as a compact manifold rather than with $\mathbb{C}^{n+1}$, and this results in greater conceptual clarity. The outlook of Chapter IV has been greatly influenced by two observations due separately to Chern and Weyl, and I take this opportunity to record them here. The first is Chern's treatment in [2] of the first main theorem using the polar divisor Σ_B as its core, and the second is the remark by Weyl in [7] that the second main theorem is nothing but a glorified version of the classical Plücker formulas. Finally, the whole of Chapter V owes such a great debt to Ahlfors' fundamental paper [1] that it would have been obvious even without my mentioning it.

Let us describe the outline of this theory in some detail. So fix a holomorphic mapping $x: V \to P_n\mathbb{C}$, where V will be assumed to carry a harmonic exhaustion as before. The first new phenomenon that appears when $n > 1$ is that x now induces a series of holomorphic mappings into various Grassmannians. To describe these, it is best to recall a well-known fact from the differential geometry of curves in $\mathbb{R}^n$. Let $\gamma: \mathbb{R} \to \mathbb{R}^n$ be a smooth curve, then the one-parameter family of tangent lines γ' (the prime denotes differentiation) along γ forms a surface, called the tangential developable of γ. Next, γ induces a one-parameter family of planes along γ which at $\gamma(t)$ is spanned by the tangent $\gamma'(t)$ and the principal normal $\gamma''(t)$. These are the osculating planes of γ, and this set of osculating planes forms a three-dimensional sub-variety of $\mathbb{R}^n$. By considering the three-dimensional osculating spaces of γ which at $\gamma(t)$ is spanned by $\gamma'(t)$, $\gamma''(t)$ and the binormal $\gamma'''(t)$, we obtain a one-parameter family of three-spaces, which forms a four-dimensional sub-variety of $\mathbb{R}^n$, and so on. Now let us return to our holomorphic curve $x: V \to P_n\mathbb{C}$. In a way that will be made precise in Chapter III, we will also be able to attach a k-dimensional <u>osculating</u> (<u>projective</u>) <u>space</u> to $x(p)$ for each $p \in V$ and for each k, $1 \leq k \leq n-1$. Thus if $G(n,k)$ denotes the Grassmannian of k-dimensional projective subspaces of $P_n\mathbb{C}$, then we can define a holomorphic map ${}_kx: V \to G(n,k)$, where ${}_kx(p)$ is just the k-dimensional osculating space at $x(p)$. The set of such k-dimensional osculating spaces ${}_kx(p)$ as p wanders through $V[t]$ forms a $(k+1)$-dimensional analytic subvariety with boundary in $P_n\mathbb{C}$, which will be denoted by $X_k(t)$. Finally, we shall agree on the notational convention ${}_0x = x$. This finite series of mappings ${}_0x, \ldots, {}_{n-1}x$, are called the <u>associated curves</u> of x.

Now let $A^k \in G(n,k)$. The set of points of $P_n\mathbb{C}$ orthogonal to A^k is an $(n-k-1)$-dimensional projective subspace of $P_n\mathbb{C}$, which is called the <u>polar space</u> of A^k and is denoted by $A^\perp$. Let $n_k(t,A^k)$ denote the intersection number of $A^\perp$ and $X_k(t)$; note that these are subvarieties of complementary dimensions, so their intersection number has an obvious meaning. Define

$$N_k(r,A^k) = \int_{r_0}^{r} n_k(t,A^k)\,dt.$$

The next observation is that $G(n,k)$ carries a classical Kahler metric called the Fubini-Study metric, whose Kahler form we denote by ω_k. Then the counterpart of the order function in the case $n = 1$ is the series of order functions:

$$T_k(r) = \frac{1}{\pi}\int_{r_0}^{r} dt \int_{V[t]} {}_kx^*\omega_k.$$

Now call a holomorphic curve <u>nondegenerate</u> if its image lies in no proper projective subspace of $P_n\mathbb{C}$. The FMT of rank k can now be stated: if x is nondegenerate, then

$$(0.9) \qquad N_k(r,A^k) + \frac{1}{2\pi}\int_{\partial V[t]} \log\frac{|_kx|}{|\langle_kx,A^k\rangle|} * d\tau\Big|_{r_0}^{r} = T_k(r),$$

where the integrand of the line integral should be interpreted in this manner: each $_kx(p)$ is a k-space, so it is representable by a unit decomposable $(k+1)$-vector, and the same is true of A^k. Since there is a well-defined inner product on $(k+1)$-vectors, $|_kx|$ and $|\langle_kx,A^k\rangle|$ make sense. The proof of (0.9) is conceptually identical with that of (0.5). However, the counterpart of (0.7) has a new element in it because it now relates three consecutive order functions T_{k-1}, T_k, T_{k+1} to the topological data of V. To state it, let $S_k(r) = \int_{r_0}^{r} s_k(t)dt$, where $s_k(t)$ denotes the <u>stationary index</u> of $_kx\colon V \to G(n,k)$ in $V[t]$, i.e. the sum of the orders of zeroes of the differential of $_kx$ in $V[t]$. Let $E(r)$ be as in (0.7). Then for any nondegenerate holomorphic curve in $P_n\mathbb{C}$, the second main theorem states that:

$$(0.10) \quad \begin{cases} E(r) + S_k(r) + (T_{k-1}(r) - 2T_k(r) + T_{k+1}(r)) \\ \qquad = \dfrac{1}{2\pi}\displaystyle\int_{\partial V[t]} \log\frac{|_{k-1}x|^2|_{k+1}x|^2}{|_kx|^4} * d\tau\Big|_{r_0}^{r} \\ \displaystyle\int_{r_0}^{r} ds \int_{r_0}^{s} \exp\{K\varphi_k(t)\}dt < CT_k(r) \end{cases}$$

where $\varphi_k(t) = \{E(t) + S_k(t) + T_{k-1}(t) - 2T_k(t) + T_{k+1}(t)\}$ and K, C are positive constants.

(0.9) and (0.10) together will lead to the defect relations. The basic idea here is still to integrate the inequality $N_0(r,a) < T_0(r) + c_0$ over $P_n\mathbb{C}$ with respect to a well-chosen function β defined on $P_n\mathbb{C}$, but the fact that $n > 1$ now makes this integration technically much more complicated than the case of $n = 1$." In a very ingenious manner, that often defies belief, Ahlfors succeeded in choosing such a β to arrive at the following inequality:

$$(0.11)\qquad (1-\alpha)\int_{r_0}^{r} dt \int_{r_0}^{t} ds \int_{\partial V[s]} \frac{|b \lrcorner_1 x|^2}{|x|^4}\left(\frac{|x|}{|\langle b,x\rangle|}\right)^{2\alpha} * d\tau < CT_0(r) + C',$$

where $0 < \alpha < 1$, b is an arbitrary unit vector and C, C' are positive constants which are independent of α and b. (Here $b \lrcorner_1 x$ stands for that vector such that if v is any other vector, $\langle b \lrcorner_1 x, v\rangle = \langle_1 x, b \wedge v\rangle$.) The delicacy of this inequality lies in the fact that $\langle b,x\rangle$ vanishes whenever $x(V)$ intersects the polar space of b (which is a hyperplane), so that the integrand has singularities in a hyperplane. It is the factor $|b \lrcorner_1 x|$ in the numerator that compensates for these singularities and prevents the integral from being divergent.

Although (0.11) is already difficult to come by, the road from (0.11) to the defect relations is still rougher. Ahlfors had to reach even greater heights in bringing this line of development to completion. The defect relations can be thus stated. For each k-dimensional projective subspace

A^k of $P_n\mathbb{C}$. let

$$\delta_k(A^k) = \lim_{r\to\infty} \inf\left(1 - \frac{N_k(r,A^k)}{T_k(r)}\right).$$

Again it is true that $0 \leq \delta_k \leq 1$. If the polar space $A^{\perp}$ of A^k never meets any k-dimensional osculating space of x, then clearly $N_k(r,A^k) = 0$ for all r and consequently $\delta_k(A^k) = 1$. Now if $x\colon V \to P_n\mathbb{C}$ is nondegenerate, V has finite Euler characteristic and $\{A^k\}$ is a system of k-spaces in general position, then

$$\sum_{A^k} \delta_k(A^k) \leq \binom{n+1}{k+1} + \chi_k, \tag{0.12}$$

where each χ_k is a finite constant and vanishes if $V = \mathbb{C}$ or $\mathbb{C} - \{0\}$. In particular, if $\{A^{n-1}\}$ is a system of $(n+2)$ hyperplanes in general position and $x\colon \mathbb{C} \to P_n\mathbb{C}$ is nondegenerate, then $x(\mathbb{C})$ must intersect one of $\{A^{n-1}\}$.

It is impossible to adequately describe the difficulty that must be surmounted in order to pass from (0.11) to (0.12). I can only refer the reader to §4 of Chapter V to fully savour this virtuoso performance of Ahlfors.

As has been remarked above, the case of $n = 1$ (i.e. meromorphic functions) suggests a lot of open problems and apparently will remain an active field for some time to come. However, the future of the general case $n > 1$ is far less certain. While these are a few obvious questions that remain unanswered (e.g. can one obtain defect relations for holomorphic curves in Grassmannians? can one replace hyperplanes by hypersurfaces of a fixed degree?*), the subject remains too narrow and too isolated and as such, it runs the risk of meeting an early and uneventful death. The most pressing problem is therefore to find applications for this theory. One such was given by Chern and Osserman in their study of minimal surfaces (J. D'Analyse Math. 1967, 15-34). In §5 of Chapter V, three more are attempted. Of these, the most interesting should be the two possible generalizations of Picard's theorem to n-dimensions. These are problems intimately connected with Kobayashi's theory of hyperbolic manifolds and ultimately with the intrinsic characterization of bounded domains. I can only hope that these notes will stimulate some interest in this subject, and that further work in this direction is forthcoming.

In conclusion, I should point out certain notational conventions employed throughout these notes:

(1) An asterisk * in front of a proof indicates that the proof can be skipped without loss of continuity.

(2) There are three distinguished functions defined on V. These are: τ (p. 32), σ (p. 35) and γ (p. 102).

(3) The sign $\|$ in front of an inequality is defined on p. 55 and p. 60.

*In a private communication, Professor Wilhelm Stoll informed me that he had solved the problem of obtaining defect relations when hyperplanes are replaced by hypersurfaces of a fixed degree. This work is unpublished.

(4) The symbol $\theta = \mu(\Phi)$ is defined on p. 131.

(5) The special usage of constants C and C' is defined on p. 168.

(6) The subscript in the symbol X_z^k is defined on p. 69 and p. 170.

(7) The special usage of $a, b, E, F,$ and A, B throughout these notes are defined on pp. 9-10.

CONTENTS

THE EQUIDISTRIBUTION THEORY OF HOLOMORPHIC CURVES

CHAPTER I

Generalities on Projective Spaces and Grassmannians

§1. We denote by $\mathbb{C}^{n+1}$ the $(n+1)$-dimensional complex euclidean space, i.e., $\mathbb{C}^{n+1} = \{(z_0, z_1, \ldots, z_n): z_A \in \mathbb{C}\}$. If $Z = (z_0, \ldots, z_n)$ and $W = (w_0, \ldots, w_n)$, then the canonical inner product on $\mathbb{C}^{n+1}$ is

$$\langle Z,W\rangle = \overline{\langle W,Z\rangle} = z_0\overline{w}_0 + \cdots + z_n\overline{w}_n.$$

Let S^{2n+1} denote the unit sphere in $\mathbb{C}^{n+1}$, i.e., $S^{2n+1} = \{Z: |Z| = 1\}$, where $|Z| \equiv +(\langle Z,Z\rangle)^{1/2}$. By $P_n\mathbb{C}$, we mean the quotient space of S^{2n+1} under the action of the circle:

$$e^{i\theta}\cdot(z_0, \ldots, z_n) = (e^{i\theta}z_0, \ldots, e^{i\theta}z_n).$$

The equivalent class of such Z we denote by $[z_0, \ldots, z_n]$. These are then the canonical principal fibre spaces:

$$\begin{array}{ccc} \mathbb{C}^{n+1} - \{0\} & \xrightarrow{\pi_1} & S^{2n+1} \\ & \pi\searrow \quad \swarrow\pi_2 & \\ & P_n\mathbb{C} & \end{array} \tag{1.1}$$

Here, $\pi(z_0, \ldots, z_n) = [\frac{z_0}{|Z|}, \ldots, \frac{z_n}{|Z|}]$, $\pi_1(z_0, \ldots, z_n) = (\frac{z_0}{|Z|}, \ldots, \frac{z_n}{|Z|})$, and $\pi_2(z_0, \ldots, z_n) = [z_0, \ldots, z_n]$. Therefore the diagram is commutative, i.e., $\pi_2 \circ \pi_1 = \pi$. The fibre of π is $\mathbb{C}^* = \mathbb{C} - \{0\}$, that of π_1 is the positive reals, and that of π_2 is the unit circle. On $\mathbb{C}^{n+1} - \{0\}$, consider the covariant 2-tensor $\tilde{F} = \frac{1}{(\Sigma_{A=0}^n z_A\overline{z}_A)^2}\{(\Sigma_A z_A\overline{z}_A)(\Sigma_A dz_A \otimes d\overline{z}_A) - (\Sigma_A\overline{z}_A dz_A)\otimes(\Sigma_A z_A d\overline{z}_A)\}$. It is hermitian, of type $(1,1)$ and in fact will be seen to be positive semi-definite. For obvious reasons, we abbreviate $\tilde{F}$ to this form:

$$(1.2) \qquad \tilde{F} = \frac{1}{\langle Z,Z\rangle^2}\{\langle Z,Z\rangle\langle dZ,dZ\rangle - \langle dZ,Z\rangle\langle Z,dZ\rangle\}$$

The associated two-form of $\tilde{F}$ will be denoted by $\tilde{\omega}$, i.e.,

$$\tilde{\omega} = \frac{\sqrt{-1}}{2}\frac{1}{(\Sigma_A z_A \bar{z}_A)^2}\{(\Sigma_A z_A \bar{z}_A)(\Sigma_A dz_A \wedge d\bar{z}_A) - (\Sigma_A \bar{z}_A dz_A) \wedge (\Sigma_A z_A d\bar{z}_A)\}$$

Again, we abbreviate $\tilde{\omega}$ to

$$(1.3) \quad \tilde{\omega} = \frac{\sqrt{-1}}{2}\frac{1}{\langle Z,Z\rangle^2}\{\langle Z,Z\rangle\langle dZ,dZ\rangle - \langle dZ,Z\rangle\langle Z,dZ\rangle\}.$$

Usually the context makes it clear whether tensor product or exterior product is intended in (1.2) or (1.3), so that there is no fear of confusion.

Consider the fibering $\pi\colon \mathbb{C}^{n+1} - \{0\} \to P_n\mathbb{C}$. This is a principal fibre bundle with $\mathbb{C}^*$ as structure group. We will now demonstrate that $P_n\mathbb{C}$ is a complex manifold, from which it will follow that π is in fact a holomorphic mapping. We will exhibit a covering of $P_n\mathbb{C}$ by complex coordinate systems. Let $U_A = \{[z_0,\ldots,z_n]\colon z_A \neq 0\}$. Then the mapping $[z_0,\ldots,z_n] \to (\frac{z_0}{z_A},\ldots,\frac{z_{A-1}}{z_A},\frac{z_{A+1}}{z_A},\ldots,\frac{z_n}{z_A})$ maps U_A bijectively onto $\mathbb{C}^n$. The open sets $U_0,\ldots,U_n$ define a complex structure on $P_n\mathbb{C}$. The following even shows that $P_n\mathbb{C}$ is a Kahler manifold.

<u>Lemma 1.1.</u> There is a Kahler metric F on $P_n\mathbb{C}$ such that $\pi^*F = \tilde{F}$. If ω is the associated Kahler form of F, then $\pi^*\omega = \tilde{\omega}$.

Remark. F is called the Fubini-Study metric of $P_n\mathbb{C}$.

Proof. The second statement is quite clear, so we will only prove the first. We will show three things to begin with: (i) $\tilde{F}$ restricted to the fibre of $\pi: \mathbb{C}^{n+1} - \{0\} \to P_n\mathbb{C}$ is zero, (ii) the action of $\mathbb{C}^*$ on $\mathbb{C}^{n+1} - \{0\}$ leaves $\tilde{F}$ invariant and (iii) the restriction of $\tilde{F}$ to the orthogonal complement of the vertical space (of $\pi: \mathbb{C}^{n+1} - \{0\} \to P_n\mathbb{C}$ with respect to the flat metric of $\mathbb{C}^{n+1}$) is positive definite.

The first two will imply the existence of such an F on $P_n\mathbb{C}$ and (iii) will imply that F is positive definite. Since (ii) is immediate from definition (1.2), we will only prove (i) and (iii). Fix a point $p = (p_0, p_1, \ldots, p_n) \in \mathbb{C}^{n+1} - \{0\}$ The vertical space at p is the complex line generated by $\Sigma_A p_A \frac{\partial}{\partial z_A} \equiv v$. (As the reader readily perceives, rather than working in the real tangent space with complex structure induced by J, we work directly with the holomorphic tangent space). Note that $|v| = |p|$. Let an orthonormal basis $\{e_0, \ldots, e_n\}$ in the tangent space to p be chosen so that $e_0 = \frac{1}{|v|} v$. Furthermore, let $e_i = \Sigma_{A=0}^n p_{Ai} \frac{\partial}{\partial z_A}$, $i = 1, \ldots, n$. The orthonormality of $e_0, \ldots, e_n$ then implies that

$$\begin{cases} \Sigma_A \, p_{Ai} \bar{p}_{Aj} = \delta_{ij} \\ \Sigma_A \, p_A \bar{p}_{Ai} = 0 \end{cases}$$

for $i, j = 1, \ldots, n$. It follows that

$$(1.4)\qquad \begin{cases} \tilde{F}(e_0, e_A) = 0, & \text{for } A = 0,\ldots,n \\ \tilde{F}(e_i, e_j) = \dfrac{1}{|p|^2}\delta_{ij}, & \text{for } i, j = 1,\ldots,n. \end{cases}$$

Consequently, the matrix of $\tilde{F}$ relative to $e_0,\ldots,e_n$ is $\frac{1}{|p|^2}\begin{bmatrix} 0 & & & \\ & 1 & & \\ & & \ddots & \\ & & & 1 \end{bmatrix}$. The fact that $\tilde{F}(e_0, e_0) = 0$ proves (i). The fact that $\tilde{F}$ restricted to the span of $\{e_1,\ldots,e_n\}$ has matrix $\frac{1}{|p|^2}\begin{bmatrix} 1 & & \\ & \ddots & \\ & & 1 \end{bmatrix}$ proves (ii).

It remains to show that F is Kahler, i.e., $d\omega = 0$. It clearly suffices to prove $d\tilde{\omega} = 0$ because π^* is injective. To this end, let

$$(1.5)\qquad d^c = \sqrt{-1}(d'' - d')$$

Then a simple computation gives

$$(1.6)\qquad \tilde{\omega} = \tfrac{1}{4} dd^c \log(\Sigma_A z_A \bar{z}_A) = \tfrac{1}{2} dd^c \log |Z|.$$

Then of course $d\tilde{\omega} = 0$. Q.E.D.

We would like to draw a consequence from the above proof. From (1.4), it follows that if $\tilde{F}$ is restricted to the unit sphere (i.e., to those p such that $|p| = 1$), its matrix relative to an orthonormal basis $\{e_0,\ldots,e_n\}$ with e_0

pointing in the radial direction is $\begin{bmatrix} 0 & & & \\ & 1 & & \\ & & \ddots & \\ & & & 1 \end{bmatrix}$. Consider the unit sphere S^{2n+1} as a real manifold and let V denote the vertical space of the fibration $\pi_1: S^{2n+1} \to P_n\mathbb{C}$. Obviously $V_a = S_a^{2n+1} \cap V'_a$, where $a \in S^{2n+1}$ and V' is the vertical space of the fibration $\pi: \mathbb{C}^{n+1} - \{0\} \to P_n\mathbb{C}$, i.e., $V'_a = \text{span}\{e_0\}$ over $\mathbb{C}$. Thus restricted to the orthogonal complement of V_a in S_a^{2n+1}, $\tilde{F}$ has matrix $\begin{bmatrix} 1 & & \\ & \ddots & \\ & & 1 \end{bmatrix}$ relative to an orthonormal basis and therefore coincides with the usual inner product on S^{2n+1}. This proves the following:

<u>Corollary</u>. If S^{2n+1} is given the usual metric and $P_n\mathbb{C}$ is given the Fubini-Study metric, then $\pi_1: S^{2n+1} \to P_n\mathbb{C}$ is an isometry when restricted to the orthogonal complement of the vertical space of this fibration.

§2. We shall need some elementary facts about Grassmannians, so we will briefly mention them. By a <u>projective subspace of dimension p in</u> $P_n\mathbb{C}$, we mean the image under $\pi: \mathbb{C}^{n+1} - \{0\} \to P_n\mathbb{C}$ of a linear subspace of $(p+1)$ dimensions in $\mathbb{C}^{n+1}$. Hence, by definition, there is a one-one correspondence between the set of linear spaces of dimension $(p+1)$ in $\mathbb{C}^{n+1}$ and the set of projective p-spaces in $P_n\mathbb{C}$ and as a rule we identify these two sets and denote this common object by $G(n,p)$. Note that $G(n,0) = P_n\mathbb{C}$. Let us fix the <u>canonical basis</u> $\epsilon_A = (0,\ldots,0,1,0,\ldots,0)$ (A-th spot) of $\mathbb{C}^{n+1}$ once and for all. $A = 0,\ldots,n$. Then we can coordinatize $G(n,p)$ as follows.

Call a $(p+1)$-vector <u>decomposable</u> iff it can be written as $e_0 \wedge \cdots \wedge e_p$. Then each $(p+1)$-dimensional subspace Λ of $\mathbb{C}^{n+1}$ defines a decomposable $(p+1)$-vector up to constant multiples. More precisely, let $\{e_0,\ldots,e_p\}$ be any basis of Λ, then

$$(1.7) \qquad e_0 \wedge \cdots \wedge e_p = \sum_{i_0<\cdots<i_p} \lambda_{i_0\cdots i_p} \epsilon_{i_0} \wedge \cdots \wedge \epsilon_{i_p}.$$

If $\{e'_0,\ldots,e'_p\}$ is a different basis of Λ, then clearly $e_0 \wedge \cdots \wedge e_p = \rho e'_0 \wedge \cdots \wedge e'_p$ for some nonzero complex number ρ. So if we write:

$$e'_0 \wedge \cdots \wedge e'_p = \sum_{i_0<\cdots<i_p} \lambda'_{i_0\cdots i_p} \epsilon_{i_0} \wedge \cdots \wedge \epsilon_{i_p},$$

then $\rho\lambda'_{i_0\cdots i_p} = \lambda_{i_0\cdots i_p}$ for all $i_0 < \cdots < i_p$. Now via the canonical basis $\{\epsilon_0,\ldots,\epsilon_n\}$, $\Lambda^{p+1}\mathbb{C}^{n+1}$ is naturally identified with $\mathbb{C}^{\ell(p)}$, where Λ^{p+1} denotes the $(p+1)$-th exterior power of $\mathbb{C}^{n+1}$ and $\ell(p) = \binom{n+1}{p+1}$. Under this identification, (1.7) implies that $e_0 \wedge \cdots \wedge e_p$ corresponds to $(\ldots,\lambda_{i_0\cdots i_p},\ldots)$ and consequently Λ itself corresponds to a line in $\mathbb{C}^{\ell(p)} - \{0\}$: $\{\rho(\ldots,\lambda_{i_0\cdots i_p},\ldots): \rho \in \mathbb{C}^*\}$. Under the projection $\pi: \mathbb{C}^{\ell(p)} - \{0\} \to P_{\ell(p)-1}\mathbb{C}$, Λ therefore corresponds to a unique point in $P_{\ell(p)-1}\mathbb{C}$. This map from $G(n,p)$ into $P_{\ell(p)-1}\mathbb{C}$ is injective (Jenner [5]). Using this map, we again identify $G(n,p)$ with its image in $P_{\ell(p)-1}\mathbb{C}$.

Before proceeding further, let us elaborate on (1.7). Suppose

$$e_a = \Sigma_{A=0}^{n} \lambda_{aA}\epsilon_A, \qquad a = 0,\ldots,p$$

then the coefficients of (1.7) satisfy

$$(1.8) \qquad \lambda_{i_0 \cdots i_p} = \det \begin{bmatrix} \lambda_{0i_0} \cdots \lambda_{0i_p} \\ \vdots \qquad \vdots \\ \lambda_{pi_0} \cdots \lambda_{pi_p} \end{bmatrix}$$

We shall also need the fact that $G(n,p)$ is a compact submanifold of $P_{\ell(p)-1}\,\mathbb{C}$. (It is in fact a quadric. Jenner [5]). Its dimension is $(n-p)(p+1)$. As we saw above, $P_{\ell(p)-1}\,\mathbb{C}$ is just the projective space formed from the vector space $\Lambda^{p+1}\mathbb{C}^{n+1}$, so this is merely the statement that the condition for a member of $\Lambda^{p+1}\mathbb{C}^{n+1}$ to be decomposable is an algebraic condition.

We should now mention that the natural inner product on $\mathbb{C}^{n+1}$ can be extended to $\Lambda^{p+1}\mathbb{C}^{n+1}$ $(-1 \leq p < \infty)$ in a natural way. If $p = -1$, $\Lambda^{0}\mathbb{C}^{n+1} = \mathbb{C}$, and the inner product is just $\langle C_1, C_2 \rangle = C_1\overline{C}_2$. Let $p \geq 0$. If $\Lambda = e_0 \wedge \cdots \wedge e_p$ and $\Xi = f_0 \wedge \cdots \wedge f_p$ are decomposable $(p+1)$-vectors, then by definition

$$(1.9) \qquad \langle \Lambda, \Xi \rangle = \det\{\langle e_a, f_b \rangle\}, \qquad a,b = 0,\ldots,p.$$

This inner product is then extended to all of $\Lambda^{p+1}\mathbb{C}^{n+1}$ by linearity. An equivalent way of describing this inner product is the following: if $\{e_0,\ldots,e_n\}$ is an orthonormal basis of $\mathbb{C}^{n+1}$, then $\{e_{i_0} \wedge \cdots \wedge e_{i_p} : i_0 < \cdots < i_p\}$ is an orthonormal basis of $\Lambda^{p+1}\mathbb{C}^{n+1}$. Yet another interpretation is this. We have identified $\Lambda^{p+1}\mathbb{C}^{n+1}$ with $\mathbb{C}^{\ell(p)}$. So let Λ, Ξ be two members of $\Lambda^{p+1}\mathbb{C}^{n+1}$ and let them correspond to λ and ξ in $\mathbb{C}^{\ell(p)}$. Then $\langle \Lambda, \Xi \rangle$ is nothing but the inner product of λ and ξ in the natural inner product of $\mathbb{C}^{\ell(p)}$. As usual, if $M \in \Lambda^{p+1}\mathbb{C}^{n+1}$, we write $|M| = +(\langle M,M\rangle)^{1/2}$.

We now relate this inner product to the differential geometry of $G(n,p)$. On $\Lambda^{p+1}\mathbb{C}^{n+1} - \{0\}$, there is the usual covariant 2-tensor corresponding to (1.2):

$$\tilde{F} = \frac{1}{\langle\Lambda,\Lambda\rangle^2}\{\langle\Lambda,\Lambda\rangle\langle d\Lambda,d\Lambda\rangle - \langle d\Lambda,\Lambda\rangle\langle\Lambda,d\Lambda\rangle\},$$

where $\Lambda = (\cdots,\lambda_{i_o\cdots i_p},\cdots)$ denotes the coordinate function of $\Lambda^{p+1}\mathbb{C}^{n+1} - \{0\}$. As before $\pi\colon \Lambda^{p+1}\mathbb{C}^{n+1} - \{0\} \to P_{\ell(p)-1}\,\mathbb{C}$ is a principal fibration, and so there is the Fubini-Study metric F on $P_{\ell(p)-1}\,\mathbb{C}$ such that $\pi^*F = \tilde{F}$. F then induces a Kahler structure on $G(n,p)$. We now show that this Kahler structure on $G(n,p)$ is homogeneous in the sense of Riemannian geometry. The unitary group $U(n+1)$ acts on projective n-space as a group of isometries with respect to the Fubini-Study metric, transitively but not effectively. This is clear from (1.2). $U(n+1)$ also acts transitively on $G(n,p)$; this is also clear because any $(p+1)$ dimensional linear space can certainly be moved to any other by the unitary group. But each unitary transformation of $\mathbb{C}^{n+1}$ induces a unitary transformation of $\Lambda^{p+1}\mathbb{C}^{n+1}$ (this is immediate from the definition (1.9)) and thus $U(n+1)$ also acts on $P_{\ell(p)-1}\,\mathbb{C}$ as a group of isometries. Combining these two facts, we see that the following holds.

Lemma 1.2 $U(n+1)$ is a transitive group of isometries on $G(n,p)$ with respect to the induced Kahler metric from $P_{\ell(p)-1}\,\mathbb{C}$.

We now come to the notion of the polar space of a projective subspace of $P_n\mathbb{C}$". If $[z_0,\dots,z_n] \in P_n\mathbb{C}$, $[w_0,\dots,w_n] \in P_n\mathbb{C}$, then the relation $\langle Z,W\rangle = 0$ is clearly independent of the particular representatives chosen. Therefore one can speak of two points of $P_n\mathbb{C}$ being orthogonal. If A is a projective subspace of dimension p, the set of points in $P_n\mathbb{C}$ orthogonal to A is clearly a projective subspace of dimension $n-p-1$, called the polar space of A, to be denoted by $A^\perp$. In particular, the polar space of a point of $P_n\mathbb{C}$ is a hyperplane, i.e., a projective $(n-1)$-space, and vice versa. Now let $B \in G(n,p)$. Up to scalar factor, $B = b_0 \wedge \cdots \wedge b_p$, where $\{b_0,\dots,b_p\}$ is a basis of B. The condition $\langle A,B\rangle = 0$ for $A \in G(n,p)$ is again independent of the particular representatives $a_0 \wedge \cdots \wedge a_p$ of A and $b_0 \wedge \cdots \wedge b_p$ of B, and the set of all $A \in G(n,p)$ satisfying $\langle A,B\rangle = 0$ will be called the polar divisor of B in $G(n,p)$ and will be denoted by Σ_B. A trivial but important remark is this: Σ_B is the intersection in $P_{\ell(p)-1}\mathbb{C}$ of $G(n,p)$ and the hyperplane defined by $\langle M,B\rangle = 0$ $(M \in \Lambda^{p+1}\mathbb{C}^{n+1})$. Consequently, Σ_B is a subvariety of $G(n,p)$ of (complex) codimension one.

There is a second interpretation of Σ_B that will be useful below:

$$\Sigma_B = \{A \in G(n,p) : A \cap B^\perp \neq \emptyset\}.$$

This follow from

Lemma 1.3 Let E and F be (p+1)-dimensional subspaces of $\mathbb{C}^{n+1}$ and let $\{e_o,\ldots,e_p\}$ and $\{f_o,\ldots,f_p\}$ be bases of E and F respectively. Then $\langle e_o \wedge \cdots \wedge e_p, f_o \wedge \cdots \wedge f_p \rangle = 0$ iff the intersection of E and the orthogonal complement of F is at least one dimensional.

To prove this, we may as well assume $\{e_o,\ldots,e_p\}$ is an orthonormal basis of E. Then extend $\{e_o,\ldots,e_p\}$ to an orthonormal basis $\{e_o,\ldots,e_n\}$ of $\mathbb{C}^{n+1}$ and express $f_o,\ldots,f_p$ in terms of $e_o,\ldots,e_n$. The lemma immediately follows.

Using this, we now give a third interpretation of Σ_B. Σ_B is in fact the generator of the $2\{(n-p)(p+1)-1\}$-dimensional homology group of $G(n,p)$. In Chern's notation [2], it is the Schubert variety $(N-p-1,\underbrace{N-p,\ldots,N-p}_{p})$ constructed from $B^{\perp} = L_{N-p-1} \subset L_{N-p+1} \subset L_{N-p+2} \subset \cdots \subset L_N = L_{N-p+p}$. This fact will be useful in Chapter IV.

§3. At this point, we wish to set up some notational conventions that will be in force throughout the rest of these notes. a, b will denote unit vectors of $\mathbb{C}^{n+1}$ as well as points of $P_n\mathbb{C}$. A, B will denote projective subspaces of $P_n\mathbb{C}$ as well as any of the unit decomposable vectors representing them. Similarly, E, F will denote linear subspaces of $\mathbb{C}^{n+1}$ as well as any of the unit decomposable vectors representing them. A superscript will indicate dimension. Thus A^h stands for both an h-dimensional projective subspace of $P_n\mathbb{C}$ and a unit decomposable (h+1)-vector representing it. Similarly, E^k will stand for both a k-dimensional linear subspace of

$\mathbb{C}^{n+1}$ and a unit decomposable k-vector representing it. G, H, K, Λ, M will be symbols for general multi-vectors without any special conventions.

Our goal in this section is to prove some elementary but useful inequalities about multi-vectors. First and easiest to prove is of course the Schwarz inequality: if $M, K \in \Lambda^{p+1}\mathbb{C}^{n+1}$, then

$$(1.10) \qquad |\langle M,K\rangle| \leq |M|\;|K|.$$

There is a more subtle inequality: let $K \in \Lambda^p\mathbb{C}^{n+1}$ and $H \in \Lambda^q\mathbb{C}^{n+1}$, then,

$$(1.11) \qquad |K \wedge H| \leq |K|\;|H|.$$

*Proof. First recall that the tensor power $\otimes^p\mathbb{C}^{n+1}$ has the following inner product: if $\{e_0,\ldots,e_n\}$ is an orthonormal basis of $\mathbb{C}^{n+1}$, then the set of $e_{i_1} \otimes \cdots \otimes e_{i_p}$ (all $i_1,\ldots,i_p$) is by definition an orthonormal basis for $\otimes^p\mathbb{C}^{n+1}$. Same for $\otimes^q\mathbb{C}^{n+1}$. Then it is routine to check that if $\kappa \in \otimes^p\mathbb{C}^{n+1}$ and $\eta \in \otimes^q\mathbb{C}^{n+1}$, $|\kappa \otimes \eta| = |\kappa|\;|\eta|$. Let $\pi\colon \otimes^s\mathbb{C}^{n+1} \to \Lambda^s\mathbb{C}^{n+1}$ be the natural projection for any s. Clearly $|\pi(\kappa)| \leq |\kappa|$ and also if $G \in \Lambda^s\mathbb{C}^{n+1}$ is given, there is a unique $\gamma \in \otimes^s\mathbb{C}^{n+1}$ such that $|\gamma| = |G|$ and $\pi(\gamma) = G$. (In fact, if $G = \Sigma\, \gamma_{i_1\cdots i_s} e_{i_1} \wedge \cdots \wedge e_{i_s}$, just take γ to be $\Sigma\, \gamma_{i_1\cdots i_s} e_{i_1} \otimes \cdots \otimes e_{i_s}$.) Now, let us pick κ and η so that $\pi(\kappa) = K$, $\pi(\eta) = H$ and $|\kappa| = |K|$, $|\eta| = |H|$. Also, since π is an algebra homomorphism, $\pi(\kappa \otimes \eta) = \pi(\kappa) \wedge \pi(\eta)$.

Thus

$$|K \wedge H| = |\pi(\kappa) \wedge \pi(\eta)| = |\pi(\kappa \otimes \eta)|$$
$$\leq |\kappa \otimes \eta| = |\kappa|\ |\eta| = |K|\ |H|. \quad \text{Q.E.D.}$$

Next we turn to the <u>interior product</u>. Let $K \in \Lambda^p\mathbb{C}^{n+1}$, $H \in \Lambda^q\mathbb{C}^{n+1}$, and assume $0 \leq p \leq q$. We define the interior product of K and H, $K \lrcorner H$, to be that $(q-p)$-vector such that, if $G \in \Lambda^{q-p}\mathbb{C}^{n+1}$,

$$\langle K \lrcorner H, G\rangle = \langle H, K \wedge G\rangle.$$

If $p = 0$, of course $K \lrcorner H = \overline{K}H$ and if $p = q$, $K \lrcorner H = \langle H,K\rangle$ by definition. On the other hand, if $p > q \geq 0$, we define $K \lrcorner H$ to be that $(p-q)$-vector such that if $M \in \Lambda^{p-q}\mathbb{C}^{n+1}$,

$$\langle M, K \lrcorner H\rangle = \langle K, H \wedge M\rangle.$$

There is a simple lemma which we shall need in Chapter III.

<u>Lemma 1.4</u>. If K and H are both decomposable multi-vectors, then $K \lrcorner H$ is also decomposable.

*<u>Proof</u>. Let $K \in \Lambda^p\mathbb{C}^{n+1}$, $H \in \Lambda^q\mathbb{C}^{n+1}$ and for definiteness, assume $p < q$. Write $\mathbb{C}^{n+1}$ as an orthogonal direct sum: $\mathbb{C}^{n+1} = H \oplus H^{\perp}$. This leads to an orthogonal projection $p: \mathbb{C}^{n+1} \to H$. If $p(K)$ is of dimension smaller than p, then K must contain an element of $H^{\perp}$. In this case, it is easy to see that $K \lrcorner H$ is zero and there is nothing to prove. So let $\dim p(K) = p$. Choose orthonormal basis $\{e_0,\ldots,e_n\}$ of $\mathbb{C}^{n+1}$, so that $\{e_0,\ldots,e_{p-1}\}$ is an orthonormal basis of $p(K)$

and so that $\{e_o,\ldots,e_{q-1}\}$ is an orthonormal basis of H. Then clearly,

$$K = \alpha e_o \wedge \cdots \wedge e_{p-1} + (\text{terms involving } e_q,\ldots,e_n).$$

It follows that $K \lrcorner H = \bar{\alpha} e_p \wedge \cdots \wedge e_{q-1}$. Q.E.D.

Our next inequality is the analogue of (1.11):

(1.12) $$|K \lrcorner H| \leq |K|\ |H|$$

This generalizes Schwarz's inequality (1.10).

*Proof. Let $K \in \Lambda^p \mathbb{C}^{n+1}$, $H \in \Lambda^q \mathbb{C}^{n+1}$ and assume $p \leq q$. If $E \in \Lambda^{q-p}\mathbb{C}^{n+1}$ is a unit $(q-p)$-vector, then (1.10) and (1.11) imply that

$$\begin{aligned} |\langle K \lrcorner H, E\rangle| &= |\langle H, K \wedge E\rangle| \\ &\leq |H|\ |K \wedge E| \leq |H|\ |K|\ |E| \\ &= |H|\ |K|. \end{aligned}$$

Taking E to be $K \lrcorner H / |K \lrcorner H|$, we are done. The proof of the case $p > q$ is similar. Q.E.D.

We close this chapter with another inequality that will be needed in Chapter V. Let E be a unit decomposable q-vector and F be a unit decomposable $(q+1)$-vector so that E is a subspace of F. Let K be a p-vector. Then

(1.13) $$|K \lrcorner E| \leq |K \lrcorner F|$$

*Proof. We first observe that if Λ and G are two decomposable q-vectors and a is a unit vector orthogonal

to Λ, then $\langle\Lambda,G\rangle = \langle\Lambda\wedge a, G\wedge a\rangle$. By our assumption, we may write $F = E\wedge a$, where a is a unit vector in F normal to E. Assume $p \le q$ and let E' be a unit $(q-p)$-vector, then

$$\langle K \,\lrcorner\, E, E'\rangle = \langle E, K\wedge E'\rangle = \langle E\wedge a, K\wedge E'\wedge a\rangle$$

$$= \langle F, K\wedge(E'\wedge a)\rangle = \langle K \,\lrcorner\, F, E'\wedge a\rangle.$$

If we let $E' = K \,\lrcorner\, E/|K \,\lrcorner\, E|$, then we obtain

$$|K \,\lrcorner\, E| = \left\langle K \,\lrcorner\, F, \frac{(K \,\lrcorner\, E)}{|K \,\lrcorner\, E|}\wedge a\right\rangle.$$

Now $\left|\frac{(K \,\lrcorner\, E)}{|K \,\lrcorner\, E|}\wedge a\right| \le 1$ because of (1.11) and since $|K \,\lrcorner\, F| = \max\limits_{|B|\le 1} \langle K \,\lrcorner\, F, B\rangle$, we have $|K \,\lrcorner\, E| \le |K \,\lrcorner\, F|$.

The proof of the case $p > q$ is similar. Q.E.D.

CHAPTER II

Nevanlinna theory of meromorphic functions

§1. Our ultimate concern is with holomorphic mappings of Riemann surfaces into $P_n\mathbb{C}$. Before taking up the general case, however, we want first to deal with the case $n = 1$, i.e., the Riemann sphere $S = P_1\mathbb{C}$, hoping thereby to gain some insight into the structure of equidistribution theory. The exposition of this chapter follows closely that of [8], and the reader may consult that paper for a more general account of Nevanlinna theory on Riemann surfaces.

In Chapter I, we defined the F-S metric (abbreviation for "Fubini-Study") on $P_n\mathbb{C}$. In the case of $S = P_1\mathbb{C}$, this simplifies to the following. π: $\mathbb{C}^2 - \{0\} \to S$ is the usual fibration, and on $\mathbb{C}^2 - \{0\}$ is the hermitian covariant tensor

$$(2.1) \qquad \tilde{F} = \frac{1}{\langle Z,Z\rangle^2}\{\langle Z,Z\rangle\langle dZ,dZ\rangle - \langle dZ,Z\rangle\langle Z,dZ\rangle\}$$

where $Z = (z_0, z_1)$ and $dZ = (dz_0, dz_1)$. The associated two-form of $\tilde{F}$ according to (1.6) is

$$(2.1)' \qquad \tilde{\omega} = \frac{1}{2}\, dd^c \log |Z| = \sqrt{-1}\; d'd'' \log |Z|$$

Let F be the F-S metric on S and let ω be its associated two-form. Then of course $\pi^*\omega = \tilde{\omega}$, $\pi^*F = \tilde{F}$ and ω is also the volume element of F on S. (In general, if in terms of a local coordinate function $z = x + \sqrt{-1}\, y$, an hermitian metric assumes the form

$$F = g(dx^2 + dy^2) = g\; dz \otimes d\bar{z},$$

then its volume element is $g\, dx \wedge dy$, while its associated two form is $\frac{\sqrt{-1}}{2} g\, dz \wedge d\bar{z}$. So they are trivially equal). Let us fix an $a \in S$. We wish to construct a function u_a on S so that $u_a(p) \to \infty$ as $p \to a$ and such that $\frac{1}{2} dd^c u_a$ on $S - \{a\}$ equals ω. Such a function was first found by Ahlfors and Shimizu in 1929, and can be described in the following way. Let (a_0, a_1) be a representative of a; (a_0, a_1) is determined up to a multiple of $e^{\sqrt{-1}\theta}$. (Recall that S is the quotient space of S^3, so $|a| = 1$). The polar space of a in S is of course just the point $a^\perp$ represented by $(-\bar{a}_1, \bar{a}_0)$, i.e., $\langle a, a^\perp \rangle = 0$. The function $|\langle Z, a^\perp \rangle|$ is unchanged if (a_0, a_1) is replaced by $(e^{\sqrt{-1}\theta} a_0, e^{\sqrt{-1}\theta} a_1)$, so we may legitimately consider it as a function of a, $a \in S$. Denote by $\tilde{u}_a(Z)$ the following function defined on $\mathbb{C}^2 - \{0\} - \{\pi^{-1}(a)\}$:

$$\log\left(\frac{|Z|}{|\langle Z, a^\perp \rangle|}\right).$$

Since $\tilde{u}_a(Z) = \tilde{u}_a(\lambda Z)$ for any nonzero complex number λ, $\tilde{u}_a$ is seen to come from a function in $S - \{a\}$, i.e., there is a unique function u_a on S, C^∞ in $S - \{a\}$, such that $\pi^* u_a = \tilde{u}_a$. Now the condition that $u_a(p) \to \infty$ as $p \to a$ may be equivalently described as $u_a(p) \to \infty$ as $\langle p, a^\perp \rangle \to 0$, and u_a is easily seen to have this property. Furthermore, $\langle Z, a^\perp \rangle$ is a holomorphic function of Z, so $d'd'' \log |\langle Z, a^\perp \rangle| = 0$. Therefore according to (2.1)', the following holds in $\mathbb{C}^2 - \{0\} - \{\pi^{-1}(a)\}$: $\frac{1}{2} dd^c \tilde{u}_a = \sqrt{-1}\, d'd'' \log \frac{|Z|}{|\langle Z, a^\perp \rangle|}$ $= \sqrt{-1}\, d'd'' \log |Z| = \tilde{\omega}$. Consequently, $\pi^*(\frac{1}{2} dd^c u_a - \omega)$

$= \frac{1}{2} dd^c \pi^* u_a - \pi^* \omega = \frac{1}{2} dd^c \tilde{u}_a - \tilde{\omega} = 0$. Since π^* is injective, $\omega = \frac{1}{2} dd^c u_a$ in $S - \{a\}$. We summarize this in the theorem below.

<u>Theorem 2.1.</u> If $a \in S$, then there is a function u_a defined on S with the following properties:

(i) u_a is C^∞ in $S - \{a\}$.

(ii) $\frac{1}{2} dd^c u_a = \omega$ in $S - \{a\}$, where ω is the volume form of the F-S metric on S.

(iii) If $\tilde{u}_a = \pi^* u_a$, then $\tilde{u}_a(Z) = \log \frac{|Z|}{|a^\perp Z|}$, where $a^\perp$ is the point in S orthogonal to a, and $\pi\colon \mathbb{C}^2 - \{0\} \to S$ is the usual fibration.

In (iii) above, we have written $|a^\perp Z|$ for $|\langle a^\perp, Z\rangle|$, <u>a convention we shall adhere to from now on.</u>

§2. To put this to use, we turn to holomorphic mappings into S. We first clarify a close relationship between holomorphic mappings into S and those into $\mathbb{C}^2$. Let V be an open Riemann surface and $x\colon V \to S$ a nonconstant holomorphic mapping. As is well-known, this is equivalent to saying that x is a meromorphic function on V. More precisely, if $x\colon V \to \mathbb{C} \cup \{\infty\}$ is a meromorphic function, let $s\colon \mathbb{C} \cup \{\infty\} \to S$ be the stereographic projection. Then $s \circ x$ is easily seen to be a holomorphic mapping into S. It is the converse of this fact that interests us, so we present it in greater detail. Let $x\colon V \to S$ be a holomorphic mapping, and we will show how it leads to a meromorphic function on V. Let $\pi: \mathbb{C}^2 - \{0\} \to S$ be the usual fibration. As we have seen in Chapter I, §1, if $U_o = \{[z_o, z_1]: z_o \neq 0\}$, then U_o is a valid coordinate neighborhood on S with complex coordinate function $\xi: U_o \to \mathbb{C}$ such that $\xi([z_o, z_1]) = z_1/z_o$. Now $S = U_o \cup \{[0,1]\}$. If V^* denotes the complement in V of

$x^{-1}([0,1])$, then $\xi \circ x \colon V^* \to \mathbb{C}$ is a holomorphic function. Let $p \in V - V^*$. We want to extend $\xi \circ x$ over $\{p\}$ so that $\xi \circ x$ becomes a meromorphic function in a neighborhood of p. To this end, let $b \in U_0$, then $b \neq x(p)$ $(= [0,1])$ and there exists a neighborhood C of $x(p)$ not containing b. Since x is holomorphic, there exists a neighborhood A of p such that $x(A) \subseteq C$. Because $V - V^*$ is discrete in V, we may assume that $A - \{p\} \subseteq V^*$. Let B be a neighborhood of b disjoint from C, then $B \cap x(A) = \emptyset$. In particular, $B \cap x(A - \{p\}) = \emptyset$. Now $\xi(B)$ is an open set in $\mathbb{C}$ because ξ is a homeomorphism of U_0 onto $\mathbb{C}$, and the one-one property of ξ implies that $\xi(B) \cap (\xi \circ x)(A - \{p\}) = \emptyset$. This means that the holomorphic function $\xi \circ x$ defined on $A - \{p\}$ omits the open set $\xi(B)$ in $\mathbb{C}$. By the theorem of Casorati-Weirstrass, p is not an essential singularity of $\xi \circ x$. This means $\xi \circ x$ can be extended to a meromorphic function $\widetilde{\xi \circ x} \colon A \to \mathbb{C} \cup \{\infty\}$. Do this for every $p \in V - V^*$ and we have extended $\xi \circ x$ to a meromorphic function $\widetilde{\xi \circ x} \colon V \to \mathbb{C} \cup \{\infty\}$. Since V is open, by the well-known theorem of Behnke-Stein, there exists a pair of holomorphic functions x_0, x_1 on V such that $\widetilde{\xi \circ x} = \dfrac{x_1}{x_0}$. Then the map $\tilde{x} \colon V \to \mathbb{C}^2$ such that $\tilde{x}(p) = (x_0(p), x_1(p))$ becomes a holomorphic mapping. Let V' be the complement in V of the common zeroes of x_0 and x_1 (which is also a discrete set), then we have the diagram:

(2.2)
$$\begin{array}{ccc} & \overset{\tilde{x}}{\nearrow} & \mathbb{C}^2 - \{0\} \\ & & \downarrow \pi \\ V' & \xrightarrow{\ x\ } & S \end{array}$$

We claim that the diagram is commutative. In order to show $\pi \circ \tilde{x} = x$ on V', it suffices to show $\pi \circ \tilde{x} = x$ on an open set, say on $V^* \cap V'$. Let $p \in V^* \cap V'$; we will prove $(\xi \circ \pi \circ \tilde{x})(p) = (\xi \circ x)(p)$. Since ξ is one-one, this will do the trick. Now, $(\xi \circ \pi \circ \tilde{x})(p) = \xi(\pi(x_0(p), x_1(p))$ $= \xi([\frac{x_0(p)}{|\tilde{x}(p)|}, \frac{x_1(p)}{|\tilde{x}(p)|}]) = \frac{x_1(p)}{x_0(p)}$, while $(\xi \circ x)(p) = \widetilde{\xi \circ x}(p)$ $= \frac{x_1(p)}{x_0(p)}$, by definition of x_0 and x_1, so we are done. We may therefore regard $x: V \to S$ as the extension of $(\pi \circ \tilde{x})|V'$ to all of V. In general, whenever diagram (2.2) is commutative, we will say that <u>the holomorphic mapping</u> $\tilde{x}: V \to \mathbb{C}^2$ <u>induces the holomorphic mapping</u> $x: V \to S$.

Conversely, suppose $\tilde{x}: V \to \mathbb{C}^2$ is a given holomorphic mapping such that $\tilde{x}(p) = (x_0(p), x_1(p))$, where x_0, x_1 are holomorphic functions, we will show that $\tilde{x}$ induces a unique holomorphic mapping $x: V \to S$ in the above sense. Let V' again be the complement of the common zeroes of x_0, x_1 in V; so $V - V'$ is a discrete set in V. $\tilde{x}$ restricted to V' gives a map $\tilde{x}: V' \to \mathbb{C}^2 - \{0\}$. Define now $x = \pi \circ \tilde{x}$. Then the diagram (2.2) is commutative by definition. If we can extend $x: V' \to S$ to be a holomorphic mapping on all of V, we will be finished. (The extension, if possible, is unique). So let $p \in V$ such that $x_0(p) = x_1(p) = 0$. Let ζ be a p-centered coordinate function in a neighborhood U of p and let

$$x_0 = a_m \zeta^m + a_{m+1} \zeta^{m+1} + \cdots$$
$$x_1 = b_n \zeta^n + b_{n+1} \zeta^{n+1} + \cdots$$

where $m, n \geq 1$ and $a_m \neq 0$, $b_n \neq 0$. We may assume U is so small that p is the only zero of x_o, x_1 in U. For definiteness, let us say $m \leq n$. Consider the new functions x_o^* and x_1^* in U such that

$$x_o^* = a_m + a_{m+1}\zeta + \cdots$$
$$x_1^* = b_n\zeta^{n-m} + b_{n+1}\zeta^{n-m+1} + \cdots$$

Since x_o only vanishes at p, x_o^* is clearly free of zeroes in U. Let $x^*: U \to \mathbb{C}^2 - \{0\}$ be such that $x^*(q) = (x_o^*(q), x_1^*(q))$. We claim: $\pi \circ x^*$ extends x from $U - \{p\}$ to U. To prove this, let again $U_o = \{[z_o, z_1]: z_o \neq 0\}$ and $\xi: U_o \to \mathbb{C}$ be $\xi([z_o, z_1]) = {z_1}/{z_o}$. We now show two things: (i) if $q \in U$ and $q \neq p$, then $(\xi \circ \pi \circ x^*)(q) = (\xi \circ x)(q)$ and (ii) $\xi \circ \pi \circ x^*$ is a holomorphic function on U. Since ξ is one-one, (i) implies that $(\pi \circ x^*) | U - \{p\} = x$ and (ii) shows that $\pi \circ x^*: U \to S$ is a holomorphic map, and we have the desired result. To prove (i), note that $q \neq p$ implies that $\zeta(q) \neq 0$. So, $(\xi \circ \pi \circ x^*)(q) = \xi(\pi((x_o^*(q), x_1^*(q)))) = \xi\left(\left[\frac{x_o^*(q)}{|x^*(q)|}, \frac{x_1^*(q)}{|x^*(q)|}\right]\right)$

$$= \frac{x_1^*(q)}{x_o^*(q)} = \frac{x_1^*(q) \cdot \zeta^m(q)}{x_o^*(q) \cdot \zeta^m(q)} = \frac{x_1(q)}{x_o(q)} = \xi(\pi(x_o(q), x_1(q))) = (\xi \circ \pi \circ \tilde{x})(q)$$

$= (\xi \circ x)(q)$ by definition of x. To prove (ii), it suffices to note that $\xi \circ \pi \circ x^* = {x_1^*}/{x_o^*}$ by the above, and since x_o^* is never zero in U, $\xi \circ \pi \circ x^*$ is evidently a holomorphic function on U. Since we can do this for each $p \in V - V'$, we have succeeded in extending $x: V' \to S$ to a holomorphic mapping $x: V \to S$.

Following Herman Weyl, we call a holomorphic mapping $x^*: U \to \mathbb{C}^2 - \{0\}$ such that $\pi \circ x^* = x$ a <u>reduced representation of</u> x <u>in</u> U. The above procedure of extension proves incidentally that if $x: V \to S$ is induced by $\tilde{x}: V \to \mathbb{C}^2$, then given any sufficiently small $U \subseteq V$, x always admits a reduced representation in U. We want to reprove this fact without the assumption that $\tilde{x}$ induces x. So let V be an <u>arbitrary</u> Riemann surface (in particular it may be compact) and let $x: V \to S$ be a holomorphic mapping. Let U be an open set of V and for definiteness, let us assume that $x(U) \subseteq U_0$, where $U_0 = \{[z_0, z_1]: z_0 \neq 0\} \subseteq S$. As usual, let $\xi([z_0, z_1]) = z_1/z_0$ be the coordinate function on U_0. Then $\xi \circ x$ becomes a holomorphic function on U and the holomorphic map $x^*: U \to \mathbb{C}^2 - \{0\}$ such that $x^*(p) = (1, (\xi \circ x)(p))$ clearly enjoys the property that $\pi \circ x^* = x$. Thus x^* is a reduced representation of x in U and we have proved that <u>if</u> $x: V \to S$ <u>is a holomorphic mapping, where</u> V <u>is arbitrary, then</u> x <u>always admits a reduced representation in any sufficiently small open set of</u> V.

§3. From this point on, let a fixed nonconstant holomorphic $x: V \to S$ be given and let D be a compact subdomain of V with C^∞ boundary. <u>In this section, we allow</u> V <u>to be arbitrary,</u> so in particular, it can be compact. If $a \in S$, we are interested in the number of pre-images of a in D, <u>counting multiplicity.</u> Now $x(p) = a$ iff $\langle x(p), a^\perp \rangle = 0$.

One can even show that the number of times p covers a is equal to the order of zero of the multi-valued holomorphic function $\langle x, a^{\perp} \rangle$ at p. Instead of proving this simple statement, we simply take as our definition:

(2.3) $n(D,a)$ = sum of the orders of zeros of
$\langle x, a^{\perp} \rangle$ in D.

We also define $v(D) = \frac{1}{\pi}\int_D x^*\omega$, where ω is the volume form of the F-S metric on S. (The reason for the factor of $\frac{1}{\pi}$ is that $\int_S \omega = \pi$; c.f. (2.10) below). The following theorem relating $n(D,a)$ to $v(D)$ is often called the non-integrated First Main Theorem.

Theorem 2.2. Assumption as above, suppose $a \in S$ is such that $f(\partial D) \cap \{a\} = \emptyset$, then

$$n(D,a) + \int_{\partial D} x^*\lambda_a = v(D) \tag{2.4}$$

where $\lambda_a = \frac{1}{2\pi} d^c u_a$. (See Theorem 2.1)

Proof. Let $x^{-1}(a) = \{a_1, \ldots, a_n\}$. By hypothesis, $\{a_1, \ldots, a_n\} \subseteq D - \partial D$, so that we can choose a neighborhood U_ϵ of a with the property that $f^{-1}(U_\epsilon) = U_1 \cup \cdots \cup U_n$, where $a_j \in U_j$, and $U_j \cap U_k = \emptyset$ if $j \neq k$. Then we have by Stokes' Theorem and Theorem 2.1 that

$$v(D) = \frac{1}{\pi}\int_D x^*\omega = \frac{1}{\pi}\lim_{\epsilon \to 0}\int_{D-U_1-\cdots-U_n} x^*\omega$$

$$= \frac{1}{\pi}\lim_{\epsilon \to 0}\int_{D-U_1-\cdots-U_n} x^*\left(\frac{1}{2}dd^c u_a\right) = \lim_{\epsilon \to 0}\int_{D-U_1-\cdots-U_n} dx^*\lambda_a$$

$$= \int_{\partial D} x^* \lambda_a + \lim_{\epsilon \to 0} \Sigma_{j=1}^n - \int_{\partial U_j} x^* \lambda_a$$

It remains to show that the last sum equals $n(D,a)$. For this purpose, it is sufficient to prove that:

(2.5) $\lim_{\epsilon \to 0} \int_{\partial U_j} -x^* \lambda_a =$ the order of zero of $\langle x(a_j), a^{\perp} \rangle$

To this end, we may clearly assume U_j is so small that we can take a reduced representation of x in U_j, i.e., let y be a holomorphic mapping $y\colon U_j \to \mathbb{C}^2 - \{0\}$ such that $\pi \circ y = x$, where $\pi\colon \mathbb{C}^2 - \{0\} \to S$. Let O.N. basis (abbreviation for "orthonormal") e_0, e_1 be chosen so that a is represented by e_0. Then $a^{\perp}$ is represented by e_1. We may also write $y = y_0 e_0 + y_1 e_1$, where y_0, y_1 are holomorphic functions. Now $(\pi^* u_a)(z) = \log \frac{|Z|}{|a^{\perp} Z|}$. So relative to e_0 and e_1, we have:

$$x^* u_a = y^* \pi^* u_a = y^* \log \frac{|Z|}{|a^{\perp} Z|}$$
$$= \log \frac{|y|}{|y_1|} = \log |y| - \log |y_1|.$$

Hence,

$$\int_{\partial U_j} -x^* \lambda_a = \frac{1}{2\pi} \int_{\partial U_j} d^c \log |y_1| - \frac{1}{2\pi} \int_{\partial U_j} d^c \log |y|.$$

Because $|y|$ is never zero in U_j, $\log |y|$ is C^∞ in U_j, so clearly $\lim_{\epsilon \to 0} \int_{\partial U_j} d^c \log |y| = 0$. To prove (2.5), it is then equivalent to proving:

$$(2.6)\qquad \lim_{\epsilon\to 0}\frac{1}{2\pi}\int_{\partial U_j} d^c \log |y_1| = \text{the order of zero of } \langle x(a_j), a^\perp\rangle$$

But this is now immediate because since $y(a_j)$ represents $x(a_j)$, we have $|x(a_j), a^\perp| = |y(a_j), a^\perp| = |y_1(a_j)|$. So if an a_j-centered coordinate function ζ is used, and if $y_1 = b_m\zeta^m + b_{m+1}\zeta^{m+1} + \cdots$, then the right side of (2.6) is of course m. On the other hand, if $h = b_m + b_{m+1}\zeta + \cdots$, then $y_1 = \zeta^m h$. Since $h(0) \neq 0$, we may assume U_j is so small that h is zero-free in U_j, so that again $\lim_{\epsilon\to 0}\int_{\partial U_j} d^c \log |h| = 0$. Hence,

$$\lim_{\epsilon\to 0}\frac{1}{2\pi}\int_{\partial U_j} d^c \log |y_1| = \frac{1}{2\pi}\lim_{\epsilon\to 0}\int_{\partial U_j} m d^c \log |\zeta|$$

$$+ \frac{1}{2\pi}\lim_{\epsilon\to 0}\int_{\partial U_j} d^c \log |h|$$

$$= \frac{m}{2\pi}\lim_{\epsilon\to 0}\int_{\partial U_j} d^c \log |\zeta| .$$

But if $\zeta = |\zeta| e^{\sqrt{-1}\theta}$, then $d^c \log |\zeta| = d\theta$. (In general, if the real functions u, v on the plane are such that $u + \sqrt{-1}\, v$ is holomorphic then a simple computation gives

$$(2.7)\qquad d^c u = dv.$$

So since $\log |\zeta| + \sqrt{-1}\,\theta$ is holomorphic, we have the above result.) Then,

$$\lim_{\epsilon\to 0}\frac{1}{2\pi}\int_{\partial U_j} d^c \log |y_1| = \frac{m}{2\pi}\lim_{\epsilon\to 0}\int_{\partial U_j} d\theta = m.$$

This proves (2.6), and hence the theorem. Q.E.D.

The nonintegrated FMT relates the important quantity $v(D)$ to the analytic quantity $n(D,a)$. We now head towards the nonintegrated Second Main Theorem, which relates $v(D)$ to the topological data on D. To state it, let D be a compact subdomain on V with C^∞ boundary ∂D as above. There is no extra work in mapping D into an arbitrary Riemann surface this time instead of the Riemann sphere S. So let $f\colon D \to M$ be a holomorphic mapping, M a compact Riemann surface. Let G be an hermitian metric on M, with Gaussian curvature K and volume form Ω. We will assume that df is nonsingular along ∂D, so that f^*G becomes an hermitian metric in a neighborhood of ∂D. Thus the geodesic curvature form κ of ∂D with respect to f^*G is defined. (A precise definition of κ is given below). Inside D however, df will in general have <u>critical points</u>, i.e., there will be a finite number of points $\{\alpha_1,\ldots,\alpha_n\} \subseteq D - \partial D$ at which $df(\alpha_j) = 0$, $j = 1,\ldots,n$. These α_j's are exactly the points which cover their images $f(\alpha_j)$ with multiplicity greater than one. If α_j covers $f(\alpha_j)$ $m(j)$-times, it is reasonable to call $(m(j)-1)$ the <u>stationary index of</u> f <u>at</u> α_j. $\Sigma_j(m(j)-1)$ will be called the <u>stationary index of</u> f <u>in</u> D <u>and will be denoted by</u> $n_1(D)$. Then the nonintegrated SMT states:

<u>Theorem 2.3</u>. Assumption and notation as above, the following is valid:

$$(2.8) \qquad 2\pi\chi(D) - \int_{\partial D}\kappa + 2\pi n_1(D) = \int_D f^*(K\Omega).$$

Furthermore, if G has constant Gaussian curvature then

$$\chi(D) - \frac{1}{2\pi}\int_{\partial D}\kappa + n_1(D) = \chi(M)v(D). \tag{2.9}$$

The χ in (2.8) and (2.9) of course denotes the Euler characteristic, and $v(D) \equiv \frac{1}{c}\int_D f^*\Omega$, where $c = \int_m \Omega$. Thus if $M = S$, this $v(D)$ coincides with our previous $v(D)$. Before proving the theorem, let us note that the F-S metric on S has constant Gaussian curvature $+4$. To see this, let $S = U_0 \cup \{[0,1]\}$ where $U_0 = \{[z_0,z_1]: z_0 \neq 0\}$. $\zeta: U_0 \to \mathbb{C}$ such that $\zeta([z_0,z_1]) = {}^{z_1}/_{z_0}$ is the usual coordinate function on U_0. Using (2.1), we find

$$F = \frac{d\zeta \otimes d\bar{\zeta}}{(1+\zeta\bar{\zeta})^2}, \qquad \omega = \frac{\sqrt{-1}}{2}\frac{d\zeta \wedge d\bar{\zeta}}{(1+\zeta\bar{\zeta})^2} \tag{2.10}$$

Now in general, if G is an hermitian metric, Ω its volume form and K its Gaussian curvature, then with respect to a local coordinate function $z = x + \sqrt{-1}\,y$, they are related thus:

$$\text{if } G = g(dx^2 + dy^2) = g\,dz \otimes d\bar{z}, \tag{2.11}$$

$$\text{then } \Omega = g\,dx \wedge dy = \frac{\sqrt{-1}}{2} g\,dz \wedge d\bar{z}$$

$$\text{and } K = -\frac{1}{2g}\Delta \log g$$

where $\Delta = \frac{\partial}{\partial x^2} + \frac{\partial}{\partial y^2} = 4\frac{\partial^2}{\partial z \partial \bar{z}}$ is the usual Laplacian. Combining (2.10) and (2.11), one readily sees that the Gaussian curvature of F is equal to 4 in U_0. Since Gaussian curvature is a continuous function, it has to be equal to 4 at $[0,1]$ also, and this proves our assertion above. Consequently,

Corollary. Let $f\colon D \to S$ be a holomorphic mapping into the Riemann sphere equipped with the F-S metric then

$$(2.12) \qquad \chi(D) + n_1(D) - \frac{1}{2\pi}\int_{\partial D} \kappa = 2v(D).$$

Theorem 2.3 will be proved via the Gauss-Bonnet Theorem. Before stating the theorem, we first give a precise definition of the geodesic curvature form of an oriented (or parametrized) curve γ in an oriented surface with respect to some Riemannian metric $\langle\,,\,\rangle$. We assume γ parametrized by arc-length s, so that ds is the volume element of γ. Let $t = d/ds$ be the unit tangent vector field along γ. Choose a unit normal field n so that $t \wedge n$ defines the same orientation as the surface i.e., if ω is the volume form of $\langle\,,\,\rangle$, then $\omega(t \wedge n) > 0$. By definition, the geodesic curvature of γ is $\kappa_g \equiv \langle D_t t, n\rangle$, where D denotes covariant differentiation. Then $\kappa \equiv \kappa_g ds$ becomes a one-form on the submanifold γ, called the geodesic curvature form of γ.

Now we quote a weak form of the Gauss-Bonnet Theorem. Let Δ be a compact Riemann surface with C^∞ boundary $\partial\Delta$ equipped with an hermitian metric G. Let K and Ω denote the Gaussian curvature and volume form of G as usual. Then the theorem states that

$$2\pi\chi(D) - \int_{\partial\Delta} \kappa = \int_{\Delta} K\Omega.$$

Two comments about this theorem are in order. First, with

respect to a local coordinate function $z = x + \sqrt{-1}\,y$, (2.11) implies that:

(2.13) if $\Omega = \frac{\sqrt{-1}}{2} g\, dz \wedge d\bar{z}$, then $K\Omega = \frac{-1}{2} dd^c \log g$.

Second, the following situation often presents itself. Δ is a compact subdomain of a Riemann surface on which is defined an hermitian metric G. Let U be a coordinate neighborhood on which $z = x + \sqrt{-1}\,y$ is valid and suppose that $\partial\Delta \cap U = \{x = \epsilon,\ \epsilon \text{ a constant}\}$ and that $\Delta \cap U = \{x \leq \epsilon\}$. Then in the notation of (2.11), we claim that the geodesic curvature form κ of $\partial\Delta$ in U has the following expression:

$$
\begin{aligned}
\kappa &= \frac{1}{2}\frac{\partial \log g}{\partial x}\, dy \qquad (2.14)\\
&= \frac{1}{2}(d^c \log g) \text{ restricted to } \partial\Delta.
\end{aligned}
$$

The equality of the last two quantities is immediate because

$$d^c \log g = -\frac{\partial \log g}{\partial y}\, dx + \frac{\partial \log g}{\partial x}\, dy,$$

while $\partial\Delta \cap U = \{x = \text{a constant}\}$. We now prove $\kappa = \frac{1}{2}\frac{\partial \log g}{\partial x}\, dy$. Since $\Delta \cap U = \{x \leq \epsilon\}$ and $G = g(dx^2 + dy^2)$, the volume form of $\partial\Delta$ is simply $\sqrt{g}\, dy$. As $\kappa = \kappa_g ds$, it suffices to prove that the geodesic curvature κ_g of $\partial\Delta$ is $\frac{1}{2\sqrt{g}}\frac{\partial \log g}{\partial x}$. Now, if t is the unit tangent vector field of $\partial\Delta$ coherent with its orientation and n is the unit normal field of $\partial\Delta$ coherent with its orientation, then clearly $t = \frac{1}{\sqrt{g}}\frac{\partial}{\partial y}$, and $n = \frac{-1}{\sqrt{g}}\frac{\partial}{\partial x}$. Let us write for simplicity $\langle\, ,\, \rangle$ for $G(\ ,\ )$. Then using $\langle \frac{\partial}{\partial x}, \frac{\partial}{\partial y} \rangle = 0$, we get:

$$\begin{aligned}
\kappa_g &= \langle D_t t, n\rangle = \frac{-1}{g}\langle D_{\frac{\partial}{\partial y}} t, \frac{\partial}{\partial x}\rangle \\
&= \frac{-1}{g}\langle \frac{1}{\sqrt{g}} D_{\frac{\partial}{\partial y}} \frac{\partial}{\partial y} + \frac{\partial}{\partial y}\left(\frac{1}{\sqrt{g}}\right)\frac{\partial}{\partial y}, \frac{\partial}{\partial x}\rangle \\
&= \frac{-1}{g\sqrt{g}}\langle D_{\frac{\partial}{\partial y}} \frac{\partial}{\partial y}, \frac{\partial}{\partial x}\rangle = \frac{1}{g\sqrt{g}}\langle \frac{\partial}{\partial y}, D_{\frac{\partial}{\partial x}} \frac{\partial}{\partial y}\rangle \\
&= \frac{1}{2g\sqrt{g}}\frac{\partial}{\partial x}\langle \frac{\partial}{\partial y}, \frac{\partial}{\partial y}\rangle = \frac{1}{2g\sqrt{g}}\frac{\partial}{\partial x} g \\
&= \frac{1}{2\sqrt{g}}\frac{\partial}{\partial x}\log g,
\end{aligned}$$

thereby proving (2.14).

Proof of Theorem 2.3. We first show how to derive (2.9) from (2.8). Suppose K is constant, then $\int_D f^*(K\Omega) = K\int_D f^*\Omega = Kcv(D)$, where it may be recalled: $v(D) = \frac{1}{c}\int_D f^*\Omega$ and $c = \int_M \Omega$. But the Gauss-Bonnet Theorem applied to M itself gives $2\pi\chi(M) = \int_M K\Omega = Kc$. Thus $\int_D f^*(K\Omega) = 2\pi\chi(M)v(D)$ and (2.9) follows.

Now to the proof of (2.8). Let $\{\alpha_1,\ldots,\alpha_m\}$ be the critical points of f and let z_j be a local coordinate function around α_j such that $z_j(\alpha_j) = 0$. Denote by W_j the set $\{|z_j| < \epsilon\}$ and let $W = \bigcup_{j=1}^m W_j$. On $D - \{\alpha_1,\ldots,\alpha_m\}$, f^*G is an hermitian metric, and so f becomes an isometry from $D - \{\alpha_1,\ldots,\alpha_m\}$ equipped with f^*G into M equipped with G. Hence $f^*(K\Omega)$ is exactly the product of the Gaussian curvature of f^*G and the volume element of f^*G in $D - \{\alpha_1,\ldots,\alpha_m\}$. So retaining the symbol κ for the geodesic curvature form of ∂W, with respect to f^*G, we apply the

Gauss-Bonnet Theorem to get:

$$2\pi\{\chi(D-W)\} - \int_{\partial D}\kappa + \int_{\partial W}\kappa = \int_{D-W} f^*(K\Omega).$$

The plus sign in front of $\int_{\partial W}\kappa$ is due to the fact that we orient each ∂W_j with respect to W_j (and not with respect to D-W). Clearly, $\chi(D-W) = \chi(D) - m$ and $\lim_{\epsilon\to 0}\int_{D-W} f^*(K\Omega) = \int_D f^*(K\Omega)$. Hence,

$$\int_D f^*(K\Omega) = 2\pi\chi(D) - \int_{\partial D}\kappa + \lim_{\epsilon\to 0}\Sigma_{j=1}^m\left(\int_{\partial W_j}\kappa - 2\pi\right)$$

To prove (2.8), and hence the theorem, it suffices to prove:

(2.15) $$\lim_{\epsilon\to 0}\left(\frac{1}{2\pi}\int_{\partial W_j}\kappa - 1\right) = \text{stationary index of } f \text{ at } \alpha_j.$$

To prove this, let us first agree to write z in place of z_j. Next, let $f(\alpha_j) = p$, and let a p-centered coordinate function ζ be chosen so that relative to ζ, the volume element Ω of G assumes the form:

$$\Omega = \frac{\sqrt{-1}}{2}\, a\, d\zeta\wedge d\bar{\zeta}.$$

Consequently,

$$f^*\Omega = \frac{\sqrt{-1}}{2}\left(f^*a\cdot\left|\frac{df^*\zeta}{dz}\right|^2\right)dz\wedge d\bar{z}$$

The stationary index of f at α_j is $(m(j)-1)$, so we may write $f^*\zeta = z^{m(j)}h(z)$, where $h(0)\neq 0$, and h is holomorphic in W_j. Then,

$$\frac{df^*\zeta}{dz} = z^{m(j)-1}\left(h(z) + z\frac{dh}{dz}(z)\right).$$

Let $b = (h + z\frac{dh}{dz})$. Since $b(0) = h(0) \neq 0$, we may as well assume W_j is so small that b is nowhere zero in W_j. Thus,

$$f^*\Omega = \frac{\sqrt{-1}}{2}(f^*a \cdot |b|^2) \cdot |z|^{2(m(j)-1)} dz \wedge d\bar{z},$$

and $(f^*a \cdot |b|^2)$ is C^∞ and zero-free in W_j. To further simplify this, let $z = |z|e^{\sqrt{-1}\,\theta}$ and let $\eta = \log |z|$. Then $\eta + \sqrt{-1}\,\theta$ is a coordinate function in W_j - {slit}, and $\partial W_j = \{\eta = \log \epsilon\}$. Now a simple computation gives $\frac{\sqrt{-1}}{2}dz \wedge d\bar{z} = |z|^2 d\eta \wedge d\theta$. Hence,

$$f^*\Omega = (f^*a \cdot |b|^2)|z|^{2m(j)} d\eta \wedge d\theta.$$

But since $f^*\Omega$ is the volume form of f^*G, the geodesic curvature form κ of ∂W_j with respect to f^*G therefore assumes the following form in accordance with (2.14): when restricted to ∂W_j,

$$\begin{aligned}\kappa &= \frac{1}{2}\, d^c \log\{f^*a \cdot |b|^2 \cdot |z|^{2m(j)}\} \\ &= m(j)\ d^c \log |z| + \frac{1}{2}\, d^c \log(f^*a \cdot |b|^2) \\ &= m(j)\ d^c\eta + \frac{1}{2}\, d^c \log(f^*a \cdot |b|^2) \\ &= m(j)d\theta + \frac{1}{2}\, d^c \log(f^*a \cdot |b|^2),\end{aligned}$$

where in the last step, we made use of (2.7). Since $f^*a \cdot |b|^2$ is nowhere zero in W_j, the last item is a C^∞ one-form in W_j. Hence,

$$\begin{aligned}\lim_{\epsilon \to 0}\left(\frac{1}{2\pi}\int_{\partial W_j}\kappa - 1\right) &= \lim_{\epsilon \to 0}\frac{m(j)}{2\pi}\int_{\partial W_j} d\theta + \lim_{\epsilon \to 0}\int_{\partial W_j}\frac{1}{4\pi}\, d^c \log(f^*a \cdot |b|)-1 \\ &= m(j) + 0 - 1 = \text{stationary index of } f \text{ at } \alpha_j.\end{aligned}$$

This proves (2.15), and hence the theorem. Q.E.D.

Before closing this section, we should point out that Theorems 2.2 and 2.3 are quite interesting when we restrict D to be a compact Riemann surface <u>without</u> boundary. So let M' be a compact Riemann surface without boundary and $f: M' \to S$ a holomorphic mapping (which, by the preceding section, is just a meromorphic function on M'). Theorem 2.2 gives $n(M',a) = v(M'$ for all $a \in S$. This is the statement that every $a \in S$ is covered exactly $v(M')$-times, thus giving the simplest instance of equidistribution. Actually, a more general statement can be proved without appeal to Theorem 2.2, namely, let $f: M' \to M$ be a holomorphic mapping between compact Riemann surfaces, then the excision axiom in homology theory readily yields that each $m \in M$ is covered exactly the same number of times (which is just the degree of f). So let us say f is n to one for some positive integer n. Then in the notation of (2.9), $v(M') = \frac{1}{c}\int_{M'} f^*\Omega = n \cdot \frac{1}{c}\int_M \Omega = n$ because $c = \int_M \Omega$. Hence (2.9) becomes:

$$\chi(M') + n_1(M') = n\chi(M),$$

a relation known as <u>Hurwitz's Formula</u>. We have only proved this on the assumption that M admits an hermitian metric of constant Gaussian curvature, but the uniformization theorem shows that this is true for all compact M, so Hurwitz's Formula holds in general. Now $n_1(M')$ is the stationary index of f in M' and is by definition nonnegative. Consequently, if $0 \geq \chi$ and $\chi(M') > \chi(M)$, there exists no nonconstant holomorphic mappings $f: M' \to M$. We shall generalize Hurwitz's Formula in Chapter IV; these are the so-called Plücker formulas.

§4. So far the open Riemann surface V has been arbitrary. To get deeper results, we must impose a restriction on V. Taking $\mathbb{C}$ as a model, which is exhaustible by the level lines of the harmonic function $\log r$, we introduce:

Definition 2.1. A C^∞ function $\tau: V \to [0,s)$ $(s \leq \infty)$ is called a harmonic exhaustion on the open Riemann surface V iff

(1) τ is onto

(2) τ is proper, i.e., inverse image of compact set is compact.

(3) τ is eventually harmonic, i.e., there exists a number $r(\tau)$, $0 \leq r(\tau) < s$, such that τ is harmonic on $\tau^{-1}([r(\tau),s))$.

As usual, a function τ is harmonic iff $dd^c\tau = 0$. With respect to a local coordinate function $z = x + \sqrt{-1}\, y$, we have $dd^c\tau = (\Delta\tau)dx \wedge dy$, where Δ is the Laplacian. So this coincides with the classical definition of harmonicity. It should be remarked that at the moment, we do not as yet distinguish between the cases $s = \infty$ and $s < \infty$, but this will be done later. If $V = M - \{a_1,\ldots,a_q\}$, where M is compact and $a_i \in M$, such a τ with $s = \infty$ can be easily constructed: choose holomorphic coordinate functions z_i about each a_i, then a C^∞ function which equals $-\log|z_i|$ in $\{|z_i| \leq \frac{1}{2}\}$

will do. A similar procedure shows that if V is obtained from a compact one by deleting a finite number of discs, such a τ with $s < \infty$ can be constructed. Finally, if V is the Riemann surface of a multivalued function such as $(\sin z)^{1/n}$, it also has a harmonic exhaustion with $s = \infty$: Let $\pi: V \to \mathbb{C}$ be the canonical projection, and let τ' be the C^∞ function on $\mathbb{C}$ which equals $\log r$ outside of the disc of radius two. Then $\tau = \tau' \circ \pi$ is obviously a harmonic exhaustion function on V. Note that this particular τ has critical points at the points whose images under π are integral multiples of π $(= 3.14159\cdots)$.

The last example shows that we must in general make allowance for τ to have many <u>critical points</u>, i.e., points at which $d\tau = 0$. However, <u>where τ is harmonic, the critical points are all isolated</u> because of the following: Let $d\tau(p) = 0$; let z be a coordinate function near p, $z = x + \sqrt{-1}\, y$. The function $f = \frac{\partial\tau}{\partial y} + \sqrt{-1}\,\frac{\partial\tau}{\partial x}$ is holomorphic because the Cauchy-Riemann equations are just $\frac{\partial^2\tau}{\partial x^2} + \frac{\partial^2\tau}{\partial y^2} = 0$, which is satisfied because τ is harmonic. Since the critical points of τ coincide with the zeroes of f, we are done.

<u>Convention</u>. From now on, we fix a harmonic exhaustion function τ on V and we work only in that part of V on which τ is harmonic. Thus all parameter values of τ which arise are always greater than $r(\tau)$.

An $r \in \mathbb{R}$ is called a <u>critical value</u> of τ iff $\tau^{-1}(r)$ contains a critical point. The above implies that all the critical values of τ in $[r(\tau), s)$ are discrete. Hence except for this discrete set of numbers, $\partial V[r]$ $(= \{p: p \in V, \tau(p) = r\})$ is an imbedded one-dimensional submanifold for all $r \geq r(\tau)$. (Since $\partial V[r]$ is compact, it has to be a disjoint union of circles). Thus $V[r]$ $(= \{p: p \in V, \tau(p) \leq r\})$ is a compact surface with C^∞ boundary except for the discrete set of critical values of τ. Furthermore, if r belongs to this exceptional discrete set, $\partial V[r]$ is a one-dimensional submanifold of V except for a finite set of points, and these are precisely the critical points of τ on $\partial V[r]$.

Before discussing the passage from τ to its associated holomorphic function σ such that $\text{Re}\,\sigma = \tau$, we digress a bit to define the conjugate differential. If φ is a one-form on V, its <u>conjugate one-form</u> $*\varphi$ is defined as follows: take any local coordinate function $z = x + \sqrt{-1}\,y$ and let $\varphi = p\,dx + q\,dy$; then by definition $*\varphi = p\,dy - q\,dx$. A simple computation will show that this definition is independent of the particular z chosen. Another simple computation will show that a real function μ is harmonic iff $d{*}d\mu = 0$. Thus returning to τ, we see that $*d\tau$ is a closed one-form on $V - V[r(\tau)]$.

Now we want to generalize the situation that in any annulus of $\mathbb{C}$, one can construct from $\log r$ the holomorphic function $\log r + \sqrt{-1}\,\theta$ (multi-valued) which gives a coordinate function on the annulus-minus-a-radial-slit. So suppose (r_1, r_2) does not contain any critical values of τ and suppose W is one component of $\text{Int}(V[r_2] - V[r_1])$. Let $\gamma = W \cap \partial V[r]$ be one of the level curves of τ in W for some $r \in (r_1, r_2)$.

We call $\int_\gamma *d\tau \equiv \Gamma$ a period of $*d\tau$ and define the conjugate harmonic function ρ of τ by $\rho(p) = \int_{p_0}^{p} *d\tau$, where p_0 is a fixed point of γ. Since $d(*d\tau) = 0$, ρ is defined up to integral multiples of Γ and $d\rho = *d\tau$. Consequently $\sigma = \tau + \sqrt{-1}\,\rho$ is a (multi-valued) holomorphic function on W. Let $\alpha = \{p \in W: \rho(p) = 0 \pmod{\Gamma}\}$ (i.e., $\rho(p) = n\Gamma$, $n \in \mathbb{Z}$). Thus α is a slit containing p_0.

<u>Lemma 2.4</u>. For $\sigma = \tau + \sqrt{-1}\,\rho$ as above, $\sigma: W - \alpha \to (r_1, r_2) \times (0, \Gamma) \subseteq \mathbb{C}$ is biholomorphic. (i.e., one-one, onto and holomorphic).

*<u>Proof</u>. Clearly σ is holomorphic, single-valued, onto and everywhere non-singular. We have to prove one-one-ness. Suppose $\sigma(p) = \sigma(p')$. Then $\tau(p) = \tau(p') = r^*$, say. Thus p and p' lie on the same level curve $\gamma^* \equiv \partial V[r^*] \cap (W-\alpha)$. If we can check that ρ is one-one on γ^*, then σ will be one-one on $W-\alpha$. Let p^* be the point of intersection of α and γ^*, then $\rho(p) = \int_{p_0}^{p} *d\tau = \int_{p_0}^{p^*} *d\tau + \int_{p^*}^{p} *d\tau$. The path of integration of the second integral may be chosen to be along γ^*. Since $d\tau$ is nowhere zero in W, $*d\tau$ is nowhere zero on γ^*. But it is easy to see that $*d\tau$ is actually nowhere zero when restricted to the submanifold γ^* so that the second integral must be either strictly increasing or strictly decreasing as p traverses γ^*. This proves that ρ is one-one on γ^*.

Q.E.D.

<u>Corollary</u>. $\int_{\partial V[r]} *d\tau \equiv L$ is a positive constant for all $r \geq r(\tau)$.

Proof. It is a continuous function of r and the lemma says that it is constant in between critical values of τ.

So L is constant. Since $*d\tau$ is coherent with the orientation of $\partial V[r]$, L is positive. Q.E.D.

Definition 2.2. Let (r_1, r_2) contain no critical value of τ. The holomorphic coordinate function of Lemma 2.4 in a component of $\tau^{-1}((r_1, r_2))$ is called a special coordinate function.

We shall always denote this special coordinate function by $\sigma = \tau + \sqrt{-1}\,\rho$. Two remarks should be made here: (i) σ is defined only up to a translation in ρ but in all applications, this ambiguity is irrelevant and will therefore be ignored. (ii) σ can be constructed locally wherever τ is free of critical points, and in the sequel we will have to invoke this local existence of σ.

Now we return to $x: V \to S$. Our immediate task is to integrate the formulas in Theorems 2.2 and 2.3. The first step in this direction is to transform the line integrals that appear in these formulas. So given $a \in S$, consider the compact surface with boundary $V[r]$ satisfying the following two conditions:

($\mathfrak{A}$) $x(\partial V[r]) \cap \{a\} = \emptyset$

($\mathfrak{B}$) r is not a critical value of τ.

($\mathfrak{B}$) implies that $\partial V[r]$ is a C^∞ submanifold of V, so according to Theorem 2.2:

$$n(r,a) + \int_{\partial V[r]} x^*\lambda_a = v(r) \tag{2.16}$$

where $\lambda_a = \frac{1}{2\pi} d^c u_a$, and we have written $n(r,a)$ for $n(V[r],a)$ and $v(r)$ for $v(V[r])$.

Lemma 2.5. Under $(\mathfrak{A})$ and $(\mathfrak{L})$, $\int_{\partial V[r]} x^*\lambda_a = \frac{\partial}{\partial r} \int_{\partial V[r]} \frac{1}{2\pi} x^* u_a * d\tau$

Proof. Since x is holomorphic, $x^*\lambda_a = \frac{1}{2\pi} d^c x^* u_a$. Since $(\mathfrak{L})$ holds, there exist r_1, r_2 such that $r \in (r_1,r_2)$ and (r_1,r_2) contains no critical values of τ. Working in one component W of $\tau^{-1}((r_1,r_2))$, we employ special coordinate function $\sigma = \tau + \sqrt{-1}\,\rho$ valid on W minus the strip $\rho = 0 \pmod{\Gamma}$. In W, $\partial V[r]$ becomes the coordinate curve $\tau = r$. Since $d^c x^* u_a = -\frac{\partial x^* u_a}{\partial \rho} d\tau + \frac{\partial x^* u_a}{\partial \tau} d\rho$, we see that $x^*\lambda_a = \frac{1}{2\pi} \frac{\partial x^* u_a}{\partial \tau} d\rho$ when restricted to $\partial V[r]$. Hence,

$$\begin{aligned}
\int_{W \cap \partial V[r]} x^*\lambda_a &= \frac{1}{2\pi} \int_{(r,0)}^{(r,\Gamma)} \frac{\partial x^* u_a}{\partial \tau} d\rho \\
&= \frac{\partial}{\partial r}\left(\frac{1}{2\pi} \int_{(r,0)}^{(r,\Gamma)} x^* u_a \, d\rho\right) \\
&= \frac{\partial}{\partial r}\left(\frac{1}{2\pi} \int_{(r,0)}^{(r,\Gamma)} x^* u_a * d\tau\right) \\
&= \frac{\partial}{\partial r}\left(\frac{1}{2\pi} \int_{W \cap \partial V[r]} x^* u_a * d\tau\right).
\end{aligned}$$

Since this is true for each component of $\partial V[r]$, we are done.

Q.E.D.

Thus if $(\mathfrak{A})$ and $(\mathfrak{L})$ hold,

$$n(r,a) + \frac{\partial}{\partial r}\left(\frac{1}{2\pi} \int_{\partial V[r]} x^* u_a * d\tau\right) = v(r)$$

We can now carry out the afore-mentioned integration: we integrate both sides with respect to r. So let $[r_1, r_2]$ be an interval in which $(\mathfrak{A})$ and $(\mathfrak{B})$ hold for all $r \in [r_1, r_2]$. Then the above leads to:

$$(2.7) \quad \int_{r_1}^{r_2} n(t,a)dt + \frac{1}{2\pi} \int_{\partial V[t]} x^* u_a * d\tau \Big|_{r_1}^{r_2} = \int_{r_1}^{r_2} v(t)dt,$$

where we have used the standard notation: $h(t)\Big|_{r_1}^{r_2} = h(r_2) - h(r_1)$. To extend (2.17) to arbitrary intervals, we need a technical lemma:

<u>Lemma 2.6.</u> $\int_{\partial V[r]} x^* u_a * d\tau$ for a fixed a is a continuous function of r for all $r \geq r(\tau)$.

We do not prove this lemma here for two reasons. (1) A more general lemma will be proved in Chapter IV. (2) We are trying to gain an understanding of holomorphic curves by looking at this special case of mapping into the Riemann sphere, so we should not be distracted by such technical details.

Now let $(r_0, r_n) \subseteq (r(\tau), s)$ (Cf. Def. 2.1). The points $x^{-1}(a)$ and the critical points of τ are both finite in the relatively compact set $\tau^{-1}((r_0, r_n))$, so the points in $[r_0, r_n]$ for which $(\mathfrak{A})$ and $(\mathfrak{B})$ do not both hold are finite in number, say, $\{r_0, r_1, \ldots, r_{n-1}, r_n\}$. We define

$$\int_{\partial V[t]} x^* u_a * d\tau \Big|_{r_i}^{r_{i+1}} = \lim_{c \uparrow r_{i+1}, d \downarrow r_i} \int_{\partial V[t]} x^* u_a * d\tau \Big|_d^c$$

Then for each (r_i, r_{i+1}), (2.17) holds and thereby giving n equations of the type (2.17). We add these n equations and using Lemma 2.6 we see that the sum of the middle terms of these equations telescopes and (2.17) becomes true for $[r_0, r_n]$ itself. In other words:

$$(2.18) \quad \int_{r_0}^{r_n} n(t,a)dt + \frac{1}{2\pi} \int_{\partial V[t]} x^* u_a *d\tau \Big|_{r_0}^{r_n} = \int_{r_0}^{r_n} v(t)dt$$

This forces the definitions:

Definition 2.3. $T(r) = \int_{r_0}^{r} v(t)dt$ is called the order function of x and $N(r,a) = \int_{r_0}^{r} n(t,a)dt$ is called the counting function.

The choice of r_0 $(r_0 \geq r(\tau))$ is immaterial so long as it is fixed once and for all. We will suppress any reference to it in the sequel. (2.18) is essentially the so-called First Main Theorem. Let us note that we can actually simplify the middle term a bit more, namely, according to diagram (2.2), $\tilde{x} = (x_0, x_1)$ and $\pi \circ \tilde{x} = x$ holds in V', where $V - V'$ is the set of the common zeroes of x_0 and x_1. For any $t \geq \tau(r)$, let $\partial V'[t] = \partial V[t] \cap V'$. $\partial V[t] - \partial V'[t]$ is then a finite set of points and so in particular is of measure zero in $\partial V[t]$. Therefore, by (iii) of Theorem 2.1:

$$\int_{\partial V[t]} x^* u_a *d\tau = \int_{\partial V'[t]} x^* u_a *d\tau = \int_{\partial V'[t]} \tilde{x}^* (\pi^* u_a) * d\tau$$

$$= \int_{\partial V'[t]} \tilde{x}^* \log \frac{|Z|}{|a^{\perp} Z|} * d\tau = \int_{\partial V'[t]} \log \frac{|\tilde{x}|}{|a^{\perp}\tilde{x}|} *d\tau$$

$$= \int_{\partial V[t]} \log \frac{|\tilde{x}|}{|a^{\perp}\tilde{x}|} *d\tau .$$

Combining this with (2.18), we have

Theorem 2.7 (FMT) For every $r \geq r(\tau)$:

$$N(r,a) + \frac{1}{2\pi} \int_{\partial V[t]} \log \frac{|\tilde{x}|}{|a^{\perp}\tilde{x}|} *d\tau \Big|_{r_0}^{r} = T(r).$$

Observe that the terms on the left depend on a whereas $T(r)$ does not. Hence the middle integral compensates for the deficiency of $N(r,a)$, (e.g., $N(r,a) = 0$ if $f(V[r]) \cap \{a\} = \emptyset$), and for this reason is sometimes referred to as the compensating term.

Now it is easy to see that, since $\partial V[t]$ is the level line of τ, $*d\tau$ induces a positive measure on $\partial V[t]$. (This is by definition of the way one orients the boundary $\partial V[t]$ of $V[t]$). Furthermore, Schwarz's inequality (1.10) implies that $\frac{|\tilde{x}|}{|a^{\perp}\tilde{x}|} \geq 1$ (one must never forget that $|a^{\perp}| = 1$), so that $\log \frac{|\tilde{x}|}{|a^{\perp}\tilde{x}|} \geq 0$. Hence,

$$(2.19) \qquad \int_{\partial V[t]} \log \frac{|\tilde{x}|}{|a^{\perp}\tilde{x}|} *d\tau > 0 \quad \text{for all } t.$$

Another technical lemma we need at this point is this:

Lemma 2.8. $\int_{\partial V[r]} \log \frac{|\tilde{x}|}{|a^{\perp}\tilde{x}|} *d\tau$ for a fixed r is a continuous function of a.

We will not prove this lemma here for the same reasons as those given after Lemma 2.6. In any case, combining these two facts, we have arrived at the following basic inequality:

(2.20) $N(r,a) < T(r) + \text{const.}$, where the constant is independent of r and a.

Proof. By Theorem 2.7,

$$N(r,a) = T(r) + \frac{1}{2\pi}\int_{\partial V[r_0]} \log \frac{|\tilde{x}|}{|a^{\perp}\tilde{x}|} *d\tau - \frac{1}{2\pi}\int_{\partial V[r]} \log \frac{|\tilde{x}|}{|a^{\perp}\tilde{x}|} *d\tau$$

$$< T(r) + \frac{1}{2\pi}\int_{\partial V[r_0]} \log \frac{|\tilde{x}|}{|a^{\perp}\tilde{x}|} *d\tau \qquad (2.19)$$

$$\leq T(r) + \text{const.}$$

where the constant is chosen to be the maximum of $\frac{1}{2\pi}\int_{\partial V[r_0]} \log \frac{|\tilde{x}|}{|a^{\perp}\tilde{x}|} *d\tau$ as a varies over the compact surface S. (Lemma 2.8). Q.E.D.

This inequality will dominate what is to follow. For the moment, however, we turn to the integration of (2.12). Let us first introduce some notation: for $x: V \to S$ as above, write $\chi(r) = \chi(V[r])$, $n_1(r) = n_1(V[r])$. Then under assumption ($\mathcal{L}$) above and

(Γ) df is nowhere zero along $\partial V[r]$,

we know from (2.12) that

$$(2.21) \qquad \chi(r) + n_1(r) - 2v(r) = \frac{1}{2\pi}\int_{\partial V[r]} \kappa$$

where κ is the geodesic curvature form of $\partial V[r]$ in the metric x^*F. (F is the F-S metric on S.) To transform the last line integral, we introduce a function h on $V - V[r(\tau)]$

(2.22) $$x^*\omega = h\, d\tau \wedge * d\tau$$

where ω is of course the volume form of the F-S metric F on S. h is not defined at the critical points of τ in $V - V[r(\tau)]$ (which is discrete), but the important thing to note is that h <u>is a non negative function</u>, for the following reason: x is holomorphic, so $x^*\omega$ is coherent with the orientation on V, while a simple computation also gives that $d\tau \wedge * d\tau$ is coherent with the orientation on V. Consequently, log h is a well-defined function on $V - V[r(\tau)]$.

<u>Lemma 2.9.</u> Under assumption ($\mathcal{L}$) and ($\sqsubset$), for every $r \geq r(\tau)$:

$$\int_{\partial V[r]} \kappa = \frac{\partial}{\partial r}\left(\frac{1}{2} \int_{\partial V[r]} (\log h) * d\tau\right).$$

<u>Proof.</u> Because of the presence of ($\mathcal{L}$) and ($\sqsubset$), log h is a C^∞ function, and because of (2.14) and (2.22),

$$\int_{\partial V[r]} \kappa = \frac{1}{2} \int_{\partial V[r]} d^c \log h$$

for every $r \geq r(\tau)$. So the lemma can evidently be proved in the same way as Lemma 2.5. Q.E.D.

The situation now parallels that of the FMT: we have by virtue of Lemma 2.9 and (2.21) that

$$\chi(r) + n_1(r) - 2v(r) = \frac{\partial}{\partial r}\left\{\frac{1}{4\pi} \int_{\partial V[r]} (\log h) * d\tau\right\}.$$

The analogue of Lemma 2.6 states that

<u>Lemma 2.10.</u> $\int_{\partial V[r]} (\log h) * d\tau$ is a continuous function

of r for all $r \geq r(\tau)$.

(This lemma will not be proved here, for the same reasons as above). So using this lemma, an integration leads to:

$$\int_{r_0}^{r} \chi(t)dt + \int_{r_0}^{r} n_1(t)dt - 2T(r) = \frac{1}{4\pi} \int_{\partial V[t]} (\log h)*d\tau \Big|_{r_0}^{r}$$

Introducing the following notation:

$$E(r) = \int_{r_0}^{r} \chi(t)dt, \qquad N_1(r) = \int_{r_0}^{r} n_1(t)dt,$$

we have arrived at the Second Main Theorem.

<u>Theorem 2.11 (SMT)</u> For $x: V \to S$ and $r \geq r(\tau)$,

$$E(r) + N_1(r) - 2T(r) = \frac{1}{4\pi} \int_{\partial V[t]} (\log h)*d\tau \Big|_{r_0}^{r}.$$

§5. The program now is to integrate the inequality (2.20) with respect to a well-chosen function on the sphere S. This is the basic idea of Nevanlinna and Ahlfors. Before doing that, however, we need to know something about integration of differential forms.

<u>Lemma 2.12.</u> Suppose $f: D \to V$ is a C^∞ map from a compact oriented manifold with boundary D into another oriented manifold V of the same dimension. Suppose Φ is an integrable form of top degree in V, and suppose for each $a \in V$, $n(a)$ denotes the <u>algebraic number</u> of preimages of a in D. Then

$$\int_D f^*\Phi = \int_M n(a)\Phi$$

Remark. The meaning of algebraic number of preimages is as follows. $n(a) = 0$ if $a \notin \operatorname{Im} f$. If $a \in \operatorname{Im} f$, then $n(a)$ is defined only when (i) $f^{-1}(a)$ is a finite number of points disjoint from ∂D, say, $\{a_1, \ldots, a_{p+q}\}$ and (ii) $df(a_j)$ is nonsingular for each j. In the event that both hold, let df preserve the orientations of D and M at p of the a_j's and reverse them at the remaining q of the a_j's. Then by definition, $n(a) = p - q$. It will be seen from the proof below that $n(a)$ is thus defined for almost all points of V. For a holomorphic f between Riemann surfaces, this definition of $n(a)$ clearly coincides with the above $n(D,a)$ except at the critical points of f (which is only finite in number and therefore has nil effect on the integration).

*Proof. Let $V' = V - \mathscr{C} - f(\partial D)$, where $\mathscr{C}$ is the image under f of the critical points of f in D (i.e., points at which df is singular). The set of critical points of f is closed in D, and because D is compact it is thus a compact set. $\mathscr{C}$ being the image of this compact set is itself compact and hence closed. Of course $f(\partial D)$ is also closed. So V' is an open submanifold of V. Furthermore, by Sard's theorem, $\mathscr{C}$ is a set of measure zero and it is well-known that $f(\partial D)$ is also a set of measure zero in V. Hence $V - V'$ has zero measure. If $v \in V'$, then $f^{-1}(v)$ is closed, and discrete (because df is nonsingular on $f^{-1}(v)$) and hence finite (because D is compact). Thus for every $v \in V'$, $n(v)$ is defined. Let $\bar{n}(v)$ be the total number of points in $f^{-1}(v)$, then each $v \in V'$ has a neighborhood in V' onto which f maps $\bar{n}(v)$ open sets diffeomorphically.

Let $\{V_j\}$ be a locally finite covering of V' by such open sets and let $\{\varphi_j\}$ be a partition of unity subordinated to $\{V_j\}$. Then $\{f^*\varphi_j\}$ is a partition of unity subordinated to the locally finite covering $\{f^{-1}(V_j)\}$ of $D-f^{-1}(\mathscr{C})-\partial D$. So by definition of the integral,

$$\Sigma_j \int_{f^{-1}(V_j)} (f^*\varphi_j)f^*\Phi = \int_{D-f^{-1}(\mathscr{C})-\partial D} f^*\Phi$$
$$= \int_D f^*\Phi,$$

where the last step is because $\int_{\partial D} f^*\Phi = 0$ and $\int_{f^{-1}(\mathscr{C})} f^*\Phi = \int_{\mathscr{C}} \Phi = 0$. Now, the definition of $n(a)$ gives:

$$\int_V n(a)\Phi = \int_{V'} n(a)\Phi$$
$$= \Sigma_j \int_{V_j} \varphi_j(n(a)\Phi)$$
$$= \Sigma_j \int_{V_j} n(a)(\varphi_j\Phi)$$
$$= \Sigma_j \int_{f^{-1}(V_j)} f^*(\varphi_j\Phi)$$
$$= \Sigma_j \int_{f^{-1}(V_j)} (f^*\varphi_j)f^*\Phi$$
$$= \int_D f^*\Phi, \qquad \text{Q.E.D.}$$

Lemma 2.12 has very interesting implications, which we now give. Recall that $N(r,a) = \int_{r_0}^{r} n(t,a)dt$, and we propose to integrate both sides relative to ω over S. In order to invert the order of integration of the right side, we shall need the following plausible fact:

Lemma 2.13. $n(t,a)$ is a measurable function on $[r_1,r_2] \times S$ where $[r_1,r_2]$ is any finite subinterval of $[0,s)$.

A more general statement will be proved in Chapter V. Now since $n(t,a) \geq 0$, Fubini's theorem gives:

$$\begin{aligned}\int_S N(r,a)\omega &= \int_{r_0}^{r} (\int_S n(t,a)\omega)dt \\ &= \int_{r_0}^{r} (\int_{V[t]} x^*\omega)dt \qquad \text{(Lemma 2.12)} \\ &= \int_{r_0}^{r} \pi v(t)dt = \pi T(r).\end{aligned}$$

In other words,

$$T(r) = \frac{1}{\pi}\int_S N(r,a)\omega. \tag{2.23}$$

Account being taken of the fact that $\int_S \omega = \pi$, we see that this says the order function is the arithmetic mean of the counting function. As a corollary, the arithmetic mean of the compensating term over S is equal to zero, a fact which also follows directly from the homogeneity of the Riemann sphere.

Armed with (2.23), we are going to show that if V carries an infinite harmonic exhaustion (i.e., $s = \infty$ in Def. 2.1), a very strong form of the Casorati-Weirstrass theorem holds on V. For $a \in S$, define

$$\delta^*(a) = \liminf_{r\to\infty} \left(1 - \frac{N(r,a)}{T(r)}\right).$$

$\delta^*(a)$ is called the <u>defect of</u> a. Before explaining the meaning of δ^*, we first show that:

$$0 \leq \delta^* \leq 1 \quad \text{on} \quad S.$$

It is obvious that $\delta^* \leq 1$ because both N and T are positive. By (2.20), $\frac{N(r,a)}{T(r)} < 1 + \frac{\text{const.}}{T(r)}$. Therefore, to show that $0 \leq \delta^*$, it suffices to show that $\lim_{r\to\infty} \frac{\text{const.}}{T(r)} = 0$, and this amounts to showing that $T(r) \to \infty$ as $r \to \infty$. But this is obvious in view of the fact that $v(t)$ is strictly increasing and positive, so that

$$T(r) = \int_{r_0}^{r} v(t)dt > v(r_0)(r-r_0) \to \infty \quad \text{as} \quad r \to \infty.$$

When is δ^* equal to 1? First of all, if $a \notin \operatorname{Im} x$, then $N(r,a) = 0$ for all r and clearly $\delta^*(a) = 1$ in this case. In general, if a is very sparsely covered by $x\colon V \to S$, so that the growth of $N(r,a)$ lags far behind that of $T(r)$, i.e., if $\limsup_{r\to\infty} \frac{N(r,a)}{T(r)} = 0$, then again $\delta^*(a) = 1$. The other extreme, when $\delta^*(a) = 0$, then obviously means that a is covered by x very often. The main observation now is that:

$$\delta^* = 0 \quad \text{almost everywhere on} \quad S.$$

This implies $x(V)$ is open dense in S, but of course it says much more. To prove this, observe that (2.23) implies that

$$\int_S \left(1 - \frac{N(r,a)}{T(r)}\right)\omega = 0.$$

On the other hand, the integrand considered as a family of functions on S depending on the parameter r is uniformly bounded because (2.20) implies that

$$1 > \left(1 - \frac{N(r,a)}{T(r)}\right) > \frac{\text{const.}}{T(r)} \to 0 \quad \text{as} \quad r \to \infty.$$

Hence by Fatou's lemma,

$$\int_S \delta^*(a)\omega = \int_S \liminf_{r\to\infty}\left(1 - \frac{N(r,a)}{T(r)}\right)\omega \le \liminf_{r\to\infty} \int_S \left(1 - \frac{N(r,a)}{T(r)}\right)\omega$$
$$= 0.$$

Since $\delta^* \geq 0$, $\int_S \delta^*(a)\omega = 0$, which is equivalent to $\delta^* = 0$ a.e.

The statement that δ^* vanishes almost everywhere on S is the essential content of the FMT; it says that the points of V are evenly distributed on all of S except for a set of measure zero. This is the first equidistribution statement we have proved thus far. Later on, we shall bring the proof of a much sharper statement: $\Sigma_{a\in S}\, \delta^*(a) \leq 2$, provided certain reasonable conditions are met.

We conclude this section with a very useful fact:

Lemma 2.14 (Concavity of the logarithm). Let μ be a positive measure on a topological space R and let E be μ-measurable. Suppose f is a nonnegative and μ-integrable function defined on E, then

$$\frac{1}{\mu(E)} \int_E (\log f)d\mu \leq \log\left\{\frac{1}{\mu(E)} \int_E f\, d\mu\right\}.$$

*Proof. The following is a trivial rephrasing of the classical proof as given in Nevanlinna [6, p. 251]. So let

$c = \frac{1}{\mu(E)} \int_E f \, d\mu$, and let $\varphi = f - c$. Note two things: (i) $\int_E \varphi \, d\mu = 0$ and (ii) $\varphi / c \geq -1$. But for $t \geq -1$, $\log(1 + t) \leq t$. Hence by (ii), $\log(1 + \varphi/c) \leq \varphi/c$, and so,

$$\frac{1}{\mu(E)} \int_E (\log f) d\mu = \frac{1}{\mu(E)} \{ \int_E \log c \, d\mu + \int_E \log(1 + \frac{\varphi}{c}) d\mu \}$$

$$\leq \log c + \frac{1}{\mu(E)} \int_E \frac{\varphi}{c} \, d\mu = \log c$$

$$\equiv \log \{ \frac{1}{\mu(E)} \int_E f \, d\mu \}. \qquad \text{Q.E.D.}$$

§6. We will now integrate $N(r,a) < T(r) + \text{const.}$ to arrive at the defect relations. Let $\rho: S \to \mathbb{R}$ be a function satisfying two conditions:

(i) it is integrable and nonnegative.

(ii) $\int_S \rho\omega = 1$.

We shall later give ρ explicitly. Since the constant in $N(r,a) < T(r) + \text{const.}$ is independent of r and a,

$$\int_S N(r,a)\rho(a)\omega < T(r) + \text{const.}$$

because of (ii). But the left side equals

$$\int_{r_0}^{r} dt \int_S n(t,a)\rho(a)\omega$$

because of (i), Lemma 2.13 and Fubini's theorem. Hence it is equal to

$$\int_{r_0}^{r} dt \int_{V[t]} x^*(\rho\omega) \qquad \text{(Lemma 2.12)}$$

$$\geq \int_{r_0}^{r} dt \int_{V[t]-V[r_0]} (x^*\rho)(x^*\omega) \qquad (x^*\rho \text{ is positive})$$

$$\cdot = \int_{r_0}^{r} dt \int_{V[t]-V[r_0]} (x^*\rho)h \; d\tau \wedge * d\tau \qquad (2.22)$$

$$= \int_{r_0}^{r} dt \int_{r_0}^{t} ds \int_{\partial V[s]} (x^*\rho)h \; * d\tau$$

where the last step may be proved by invoking the special coordinate function $\sigma = \tau + \sqrt{-1}\,\rho$. Hence,

$$(2.24) \qquad \int_{r_0}^{r} dt \int_{r_0}^{t} ds\left(\int_{\partial V[s]} (x^*\rho)h \; * d\tau\right) < T(r) + \text{const.}$$

We now apply Lemma 2.14 and the Corollary to Lemma 2.4:

$$\frac{1}{L} \int_{\partial V[s]} \log\{(x^*\rho)h\} * d\tau \leq \log\left\{\frac{1}{L} \int_{\partial V[s]} (x^*\rho)h \; * d\tau\right\}$$

or,

$$\log L + \frac{1}{L}\left\{\int_{\partial V[s]} \log x^*\rho \; * d\tau + \int_{\partial V[s]} \log h \; * d\tau\right\}$$

$$\leq \log \int_{\partial V[s]} (x^*\rho)h \; * d\tau$$

So by (2.24):

$$(2.25) \qquad \int_{r_0}^{r} dt \int_{r_0}^{t} \exp\left\{\log L + \frac{1}{L} \int_{\partial V[s]} \log(x^*\rho) * d\tau + \frac{1}{L} \int_{\partial V[s]} \log h \; * d\tau\right\} d$$

$$< T(r) + \text{const.}$$

We wish to bring $\int_{\partial V[s]} \log(x^*\rho) * d\tau$ to the form of the line integral in the FMT. By the argument preceding Theorem 2.7, $x^*\rho$ should be $\frac{|\tilde{x}|}{|a^{\perp}\tilde{x}|}$. More correctly, let $\{a_1, \ldots, a_q\}$ be a finite set of distinct points of S. Consider the following function $\tilde{\rho}$ on $\mathbb{C}^2 - \{0\}$:

$$\tilde{\rho}(Z) = c \prod_{i=1}^{q} \left(\frac{|Z|}{|a_i^{\perp} Z|}\right)^{2\lambda}$$

where c is some positive constant to be determined, and $0 < \lambda < 1$. As usual, the fact that $\tilde{\rho}(eZ) = \tilde{\rho}(Z)$ for all $e \in C^*$ implies the existence of a function ρ on S such that $\pi^* \rho = \tilde{\rho}$, where $\pi\colon \mathbb{C}^2 - \{0\} \to S$. We first check that ρ is an integrable function on S. The only places that may cause trouble are exactly the points $\{a_1, \ldots, a_q\}$. Let us fix an a_j and inspect ρ near a_j. Now if $i \neq j$, $|a_i^{\perp} a_j| > 0$ and so $\frac{|Z|}{|a_i^{\perp} Z|}$ is continuous near a_j. Thus we need only concentrate our attention on the factor $\frac{|Z|}{|a_j^{\perp} Z|}$. Choose O.N. basis e_0, e_1 in $\mathbb{C}^2$ so that a_j is represented by e_0, then $a_j^{\perp}$ is represented by e_1. Let $\zeta\colon U_0 \to \mathbb{C}$ be the coordinate function given by $\zeta([z_0 e_0 + z_1 e_1]) = {}^{z_1}/_{z_0}$. Clearly ζ is an a_j-centered coordinate system. Since

$$\left(\frac{|Z|}{|a_j^{\perp} Z|}\right)^{2\lambda} = \left(\frac{z_0 \bar{z}_0 + z_1 \bar{z}_1}{z_1 \bar{z}_1}\right)^{\lambda} = \left(1 + \left|\frac{z_1}{z_0}\right|^{-2}\right)^{\lambda},$$

we see that with respect to the coordinate function ζ, $\left(\frac{|Z|}{|a_j^{\perp} Z|}\right)^{2\lambda}$ assumes the form: $(1 + |\zeta|^{-2})^{\lambda}$. It is consequently equal to

$$(|\zeta|^2 + 1)^{\lambda} |\zeta|^{-2\lambda} = \text{continuous function} \cdot |\zeta|^{-2\lambda}.$$

Since $0 < \lambda < 1$, polar coordinates in the plane clearly says that $|\zeta|^{-2\lambda}$ is integrable near the origin, i.e., $\left(\frac{|Z|}{|a_j^{\perp} Z|}\right)^{2\lambda}$

is integrable near a_j. Since this holds for each j, ρ is integrable on S. As ρ is clearly nonnegative, condition (i) of the requirement of ρ is met. To cope with (ii), we simply make an appropriate choice of c.

Thus we have chosen our ρ. The usual arguments then give:

$$\int_{\partial V[s]} \log(x^* \rho) * d\tau = \int_{\partial V[s]} \log \tilde{x}^* (\pi^* \rho) * d\tau = \int_{\partial V[s]} \log(\tilde{x}^* \tilde{\rho}) * d\tau$$

$$= 2\lambda \, \Sigma_{i=1}^{q} \int_{\partial V[s]} \log \frac{|\tilde{x}|}{|a_i^{\perp} \tilde{x}|} * d\tau + \text{const.}$$

Combining this with (2.25) and letting $\lambda \uparrow 1$, we obtain:

$$\int_{r_0}^{r} dt \int_{r_0}^{t} \exp\{\text{const.} + \frac{2}{L} \Sigma_{i=1}^{q} \int_{\partial V[s]} \log \frac{|\tilde{x}|}{|a_i^{\perp}\tilde{x}|} * d\tau$$

$$+ \frac{1}{L} \int_{\partial V[s]} \log h * d\tau\} ds$$

$$\leq T(r) + \text{const.}$$

Substituting into this inequality the expression of the line integrals in the FMT and SMT, we have clearly proven the following: if we define

$$\varphi(s) = \frac{4\pi}{L}\{\Sigma_{k=1}^{q}(T(s) - N(a_k,s)) + N_1(s) - 2T(s) + E(s)\}$$

$$+ \text{const.},$$

where the constant is independent of $\{a_1,\ldots,a_q\}$ and s, then

$$(2.26) \qquad \int_{r_0}^{r} dt \int_{r_0}^{t} e^{\varphi(s)} ds \leq T(r) + \text{const.}$$

We now bring to (2.26) a minor refinement. Define $\overline{n}(r,a)$ to be the number of zeroes of $\langle x, a^{\perp} \rangle$ in $V[r]$ <u>without</u> counting multiplicity, and let

$$\overline{N}(r,a) = \int_{r_o}^{r} \overline{n}(t,a)dt.$$

We clearly have: $n(r,a) - \overline{n}(r,a) = \Sigma_{p \in x^{-1}(a) \cap V[r]}$ (stationary index of x at p) ≥ 0, so that $n_1(r) \equiv \Sigma_{p \in V[r]}$ (stationary index of x at p) $\geq \Sigma_{k=1}^{q} \{n(r,a_k) - \overline{n}(r,a_k)\} \geq 0$. Hence, $N_1(r) \geq \Sigma_{k=1}^{q}(N(r,a_k) - \overline{N}(r,a_k))$, which implies that

$$\Sigma_k(T(r) - N(r,a_k)) + N_1(r) \geq \Sigma_k(T(r) - \overline{N}(r,a_k)).$$

Combining this with (2.26), we have the final result:

<u>Theorem 2.15</u>. If $\varphi(r)$ denotes the quantity:

$$\frac{4\pi}{L}\{\Sigma_{k=1}^{q}(T(r) - \overline{N}(r,a_k)) - 2T(r) + E(r)\} + \text{const.}$$

then

$$\int_{r_o}^{r} dt \int_{r_o}^{t} e^{\varphi(s)} ds < T(r) + \text{const.},$$

where the constants are independent of r and $\{a_1, \ldots, a_q\}$.

From Theorem 2.15, we are going to derive bounds of $\varphi(r)$ directly in terms of $T(r)$. We seek inequality of the type: $e^{\varphi(r)} \leq (T(r) + \text{const.})$. Naturally, we must make allowance for the fact that such an inequality may not hold for <u>all</u> r. To this end, we distinguish two cases.

<u>Definition 2.4</u>. V is said to admit an <u>infinite harmonic exhaustion</u> if $s = \infty$ in Definition 2.1. If $s < \infty$, we say

V admits a finite harmonic exhaustion.

CASE 1. The infinite case.

Lemma 2.16. If ψ is a once continuously differentiable, positive, increasing function on $[r_o,\infty)$ with $r_o \geq 1$, then for any real number $k > 1$, $\psi' \leq \psi^k$ holds on $[r_o,\infty) - I$, where I is an open set on which $\int_I d \log x < \infty$. (I depends on k.)

Proof. Let $I \subseteq [r_o,\infty)$ be the open set on which $\psi' > \nu\psi^k$ where ν is any positive function and $k > 1$ is an arbitrary positive number. Then,

$$\nu < \frac{1}{\psi^k}\psi' = \frac{1}{(1-k)}(\psi^{1-k})' \quad \text{on } I,$$

$$\Longrightarrow \int_I \nu < \int_I \frac{1}{1-k}(\psi^{1-k})' = \Sigma_i \left.\frac{1}{(1-k)\psi^{k-1}}\right|_{a_i}^{b_i}$$

$$\leq \lim_{i\to\infty} \frac{1}{1-k}\left(\frac{1}{\psi^{k-1}(b_i)} - \frac{1}{\psi^{k-1}(a_1)}\right) < \infty$$

where we have written $I = \bigcup_i (a_i,b_i)$ and the last inequality is because ψ is positive and increasing. So outside of I, $\psi' \leq \nu\psi^k$. If we let $\nu = \frac{1}{x}$, $r_o \geq 1$ implies $\nu \leq 1$ on $[r_o,\infty)$ and the lemma is now obvious. Q.E.D.

Successive applications of Lemma 2.16 to $\psi(t) = \int_{r_o}^t e^{\varphi(s)}ds$ and $\psi(r) = \int_{r_o}^r dt \int_{r_o}^t e^{\varphi(s)}ds$ yield: if $k > 1$, then $e^{\varphi(t)} \leq (\int_{r_o}^t e^{\varphi(s)}ds)^k$ for $t \notin I'$ such that $\int_{I'} d \log x < \infty$; $\int_{r_o}^t e^{\varphi(s)}ds \leq (T(t) + \text{const.})^k$ for $t \notin I''$ such that $\int_{I''} d \log x < \infty$. The latter made use of Theorem 2.15. Let

$I = I' \cup I''$, then

$$e^{\varphi(r)} \leq (T(r) + \text{const.})^{k^2}$$

for $r \notin I$ such that $\int_I d \log x < \infty$. In general, we shall let "$\|$" stand in front of an inequality if the latter only holds in $[r_o, \infty) - I$. Thus,

$$\|\varphi(r) < k^2 \log(T(r) + \text{const.}),$$

$$\|\Sigma_{k=1}^{q}(T(r) - \overline{N}(r, a_k)) \leq 2T(r) - E(r) + \frac{Lk^2}{4\pi} \log(T(r) + \text{const.} \\ + \text{const.}$$

by virtue of the expression of $\varphi(r)$ in Theorem 2.15. Equivalently,

$$\|\Sigma_k(1 - \frac{\overline{N}(r, a_k)}{T(r)}) \leq 2 + \frac{-E(r)}{T(r)} + \frac{1}{T(r)}\{\frac{Lk^2}{4\pi} \log(T(r) + \text{const.}) \\ + \text{const.}\}.$$

The important thing to observe is that since $\int_I d \log x < \infty$, I does not comprise a neighborhood of ∞, so that the above inequality holds in an unbounded set. We are therefore prompted to introduce:

$$\delta(a) = \liminf_{r \to \infty}(1 - \frac{\overline{N}(r, a)}{T(r)})$$

$$\chi = \limsup_{r \to \infty}(\frac{-E(r)}{T(r)}).$$

The above inequality therefore reads:

$$\Sigma_{k=1}^{q} \delta(a_k) \leq 2 + \chi + \limsup_{r \to \infty} \{ \}.$$

Since $T(r) \to \infty$ as $r \to \infty$ and since the constants in { } are all independent of r and $\{a_1,\ldots,a_q\}$, l'Hôpital's rule implies that the last term vanishes. Consequently,

<u>Theorem 2.17 (Defect Relations)</u>. If $x: V \to S$ is holomorphic and if V admits an infinite harmonic exhaustion, then for any finite set of distinct points $\{a_1,\ldots,a_q\}$ of S,

$$\Sigma_{k=1}^{q} \delta(a_k) \leq 2 + \chi \tag{2.27}$$

δ, like δ^* introduced in §5 above, is also called the <u>defect</u> function. Since $N(r,a) \geq \overline{N}(r,a)$, it is clear that $\delta^* \leq \delta$. Hence $0 \leq \delta \leq 1$ again holds. $\delta(a) > 0$ iff it is lightly covered compared with the other points. If $\delta(a) = 1$, we say a is a <u>Picard point</u>. If $a \in S - x(V)$, surely a is a Picard point, but the converse is by no means true (e.g. ze^z assumes the value 0, but $\delta(0) = 1$). The reader will easily convince himself that from (2.27) follows the relation:

$$\Sigma_{a \in S} \delta(a) \leq 2 + \chi.$$

For when χ is finite, $\delta \equiv 0$ except on a countable set. We now elaborate on χ. It evidently measures the relative growth of the Euler characteristic of $V[r]$ against the growth of the order function. If the number of components of $\partial V[r]$ does not stay constant for large r, then $\chi(V[r]) \to \infty$ as $r \to \infty$. Hence in this case, the usual direct-limit argument with homology groups gives $\chi(V) = -\infty$. Thus if $\chi(V) > -\infty$, there is an r_o so that $r > r_o \Longrightarrow \chi(V[r])$ is a fixed constant (equal to $\chi(V)$). Therefore, when $\chi(V) \geq 0$, i.e., if V is $\mathbb{C}$ or $\mathbb{C} - \{0\}$, $\chi \leq 0$ so that (2.27) reduces to $\Sigma_{k=1}^{q} \delta(a_k) \leq 2$. This is a far-reaching generalization of Picard's Theorem. In general, $\chi > 0$ is to be expected because we may delete as many points as we wish from S to obtain an cpen Riemann surface S' which admits an infinite harmonic exhaustion; the natural injection of $S' \to S$ certainly cannot obey any defect relation of the type $\Sigma_k \delta(a_k) \leq 2$. So we should seek a condition on x itself to insure the vanishing of χ. Here is one. We call $x: V \to S$ <u>transcendental</u> iff $\lim_{r\to\infty} \frac{r}{T(r)} = 0$. Then one can easily prove that if x is transcendental and $\chi(V)$ is finite, then $\chi = 0$. In fact let r_o be so large that $\chi(V[r]) = \chi(V)$ for all $r \geq r_o$. Then,

$$\chi = \lim\sup \frac{-E(r)}{T(r)} = \lim\sup \frac{-1}{T(r)} \int_{r_o}^{r} \chi(t)dt$$

$$= \lim\sup \frac{-1}{T(r)} \cdot \chi(V)(r-r_o) = 0$$

In a special case, the notion of transcendency coincides with the classical notion of essential singularity. For there is this result:

<u>Lemma 2.18.</u> If V is obtained from a compact Riemann surface M by deleting a finite number of points $\{a_1,\ldots,a_m\}$, then $x: V \to S$ is transcendental iff x is not extendable to a holomorphic mapping $x': M \to S$.

Proof. If $x: V \to S$ is transcendental, we will first show that it is not extendable to $x': M \to S$. In fact, we prove a more general statement: if V admits an infinite harmonic exhaustion and $x: V \to S$ is transcendental, then for every real number $r \in x(V - V[r])$ is dense in S.

Suppose false, then there exists an $a \in S$ and a neighborhood U of a such that $U \cap x(V - V[r]) = \emptyset$ for some $r \in \mathbb{R}$. There is no harm in letting $r = r_o$. So $n(t,a) = n(r_o,a)$ for all $t \geq r_o$, Hence,

$$N(r,a) = \int_{r_o}^{r} n(t,a)dt = n(r_o,a)(r-r_o).$$

Next, since $t \geq r_o \Longrightarrow x(\partial V[t]) \cap U = \emptyset$, x^*u_a restricted to $\partial V[t]$ for all such t is bounded above by a constant K. (See Theorem 2.1) By (2.18),

$$\begin{aligned} T(r) &= N(r,a) + \frac{1}{2\pi} \int_{\partial V[t]} x^* u_a * d\tau \Big|_{r_o}^{r} \\ &\leq N(r,a) + \frac{1}{2\pi} \int_{\partial V[r]} x^* u_a * d\tau \\ &\leq n(r_o,a)(r-r_o) + \frac{K}{2\pi} \cdot L \end{aligned}$$

by the corollary to Lemma 2.4. So clearly, $\limsup \frac{r}{T(r)} \geq \frac{1}{n(r_o,a)} > 0$. This contradicts transcendency.

We now prove the converse: if x is not transcendental, then it is extendable to an $x': M \to S$. So let $\limsup \frac{r}{T(r)} = \beta > 0$. Then $\liminf \frac{T(r)}{r} = \frac{1}{\beta} < \infty$. Since $N(r,a) < T(r) + \text{const}$ for all $a \in S$ ((2.20)), $\liminf \frac{N(r,a)}{r} < \frac{1}{\beta} < \infty$. But

$$n(r,a) = \frac{\frac{d}{dr} N(r,a)}{\frac{d}{dr} r},$$

and $n(r,a)$ has a limit as $r \to \infty$ because it is monotone increasing, so $\lim \frac{N(r,a)}{r}$ exists and equals $\lim n(r,a)$ by l'Hôpital's rule. Hence $\lim n(r,a) < \frac{1}{\beta} < \infty$. But the number β is independent of a, so the number of pre-images of a for all $a \in S$ is bounded by a universal constant.

Now let U_0 be the open set in S such that $U_0 = \{[z_0,z_1]: z_0 \neq 0\}$ and let $\xi: U_0 \to \mathbb{C}$ be the usual coordinate function $\xi([z_0,z_1]) = z_1/z_0$. Then $S = U_0 \cup \{[0,1]\}$. After the reasoning of §2, $x: V \to S$ is holomorphic iff $\widetilde{\xi \circ x}: V \to \mathbb{C} \cup \{\infty\}$ is a meromorphic function. According to the preceding paragraph, this meromorphic function $\widetilde{\xi \circ x}$ has the property that its preimage of any member of the extended complex plane is a finite number of points. But $V = M - \{a_1,\ldots,a_m\}$, so if $\widetilde{\xi \circ x}$ has an essential singularity at any a_j, the Casorati-Weirstrass theorem coupled with the Baire category theorem would imply that there is at least one $a \in \mathbb{C}$ whose preimage is an infinite set. This not being the case, $\widetilde{\xi \circ x}$ is extendable to a meromorphic function on M, and consequently x itself is extendable to a holomorphic mapping $x': M \to S$.

Q.E.D.

For further applications and examples concerning defect values, points of ramification and uniqueness theorems, the reader is referred to Nevanlinna [6], Hayman [4], [8], as well as a forthcoming dissertation by Edwardine Schmid (Berkeley 1969).

CASE 2. The finite case.

Lemma 2.19. Suppose ψ is a once continuous differentiable positive increasing function on $[0,s)$, $s < \infty$. Then for any real number $k > 1$, $\psi'(r) \leq \frac{1}{s-r}\psi^k(r)$ holds for all $r \in [0,s) - I$, where I is an open set on which $\int_I d\log(s-r) > -\infty$.

The proof of this theorem is analogous to that of Lemma 2.16. As before, the main thing we should note is that I does not comprise a full neighborhood of s in $[0,s)$ so that $\psi'(r) \leq \frac{1}{s-r}\psi^k(r)$ is true for numbers arbitrarily close to s. By applying this lemma successively to $\int_{r_0}^r e^{\varphi(s)}ds$ and $\int_{r_0}^r dt\int_{r_0}^t e^{\varphi(s)}ds$, we obtain:

$$\|\varphi(r) \leq k^2\log(T(r) + \text{const.}) + (k+1)\log(\frac{1}{s-r})$$

where "$\|$" now means "outside of I". This together with Theorem 2.15 imply that

$$\begin{aligned}\|\Sigma_k(1 - \frac{\overline{N}(r,a_k)}{T(r)}) &\leq 2 + \frac{-E(r)}{T(r)} + \frac{1}{T(r)}\{\frac{Lk^2}{4\pi}\log(T(r) + \text{const.}) \\ &\qquad + (k+1)\log\frac{1}{s-r} + \text{const.}\} \\ &\equiv 2 + \frac{-E(r)}{T(r)} + \epsilon(r).\end{aligned}$$

So we introduce again:

$$\delta(a) = \liminf_{r\to s}(1 - \frac{\overline{N}(r,a)}{T(r)})$$

$$\chi = \limsup_{r\to s}\frac{-E(r)}{T(r)}$$

$$\epsilon = \limsup_{r\to s}\frac{1}{T(r)}\{c_1\log(T(r) + c_2) + 2\log\frac{1}{s-r} + c_3\},$$

where c_1, c_2, c_3 are some positive constants. The above inequality leads to

<u>Theorem 2.20 (Defect relations)</u>. If $x: V \to S$ is holomorphic and V admits a finite harmonic exhaustion, then for every finite set of distinct points $\{a_1, \ldots, a_q\}$ of S:

$$\Sigma_{k=1}^{q} \delta(a_k) \leq 2 + \chi + \epsilon.$$

Furthermore $\epsilon = 0$ if

$$\limsup_{r \to s} \frac{\log \frac{1}{s-r}}{T(r)} = 0. \qquad (2.28)$$

If (2.28) holds and $\chi(V) > -\infty$, it is obvious that $\chi = 0$ also. So in this case, $\Sigma_k \delta(a_k) \leq 2$ again holds. It is not clear what the exact meaning of (2.28) is. For V = unit disc, one can show that if x has a very bad essential singularity on the unit circle, then (2.28) is satisfied. Yet, there are meromorphic functions with an essential singularity on the unit circle for which (2.28) does not hold. In general, one can show that if (2.28) does hold, then $x(V - V[r])$ is dense in S in a very strong sense. See [8] for details. The other extreme of (2.28) is the case where $T(r)$ is bounded. For such meromorphic functions defined on the unit disc, there is a vast literature. One classical theorem due to Nevanlinna is that such a meromorphic function is the quotient of two bounded holomorphic functions. The reader should consult Hayman [4] for this and related matters.

CHAPTER III

Elementary properties of holomorphic curves

§1. Our object of study is a holomorphic curve, i.e., an $x: V \to P_n\mathbb{C}$, where x is a holomorphic map and V is an open Riemann surface. The most natural way to generate such a map is given by a system of $(n+1)$ holomorphic functions $(x_0, \ldots, x_n)$ not all of them identically zero; in other words, a holomorphic map $\tilde{x}: V \to \mathbb{C}^{n+1}$ is given such that $\tilde{x} = (x_0, \ldots, x_n)$ and the vector valued function $\tilde{x}$ is not identically zero. We now elaborate on this. Let V' be the complement of the common zeroes of $\{x_0, \ldots, x_n\}$ in V, then we have the following commutative diagram:

$$\begin{array}{ccc} & & \mathbb{C}^{n+1} - \{0\} \\ & \overset{\tilde{x}}{\nearrow} & \big\downarrow \pi \\ V' & \xrightarrow{\quad x \quad} & P_n\mathbb{C} \end{array} \tag{3.1}$$

where by definition, $x = \pi \circ \tilde{x}$. In a moment, we will extend x to be a holomorphic map on all of V. Whenever we have a holomorphic $x: V \to P_n\mathbb{C}$ and $\tilde{x}: V \to \mathbb{C}^{n+1}$ such that (3.1) is commutative, we say $\tilde{x}$ induces x. The extension of x to V is achieved in the following manner. Let $p \in V - V'$, then $x_0(p) = \cdots = x_n(p) = 0$. Since $V - V'$ is discrete, we may choose a coordinate function z centered at p defined in a neighborhood U of p such that $U - \{p\} \subseteq V'$. In U, we have the following factorizations:

$$x_A(z) = z^{\delta_A} y_A(z), \qquad A = 0, \ldots, n,$$

where $\delta_A \geq 1$, y_A is holomorphic in U and $y_A(0) \neq 0$. We may assume U is so small that y_A is zero-free in all of U for every A. Among the integers $\delta_0, \ldots, \delta_n$, there is a smallest one; there is no harm in assuming that it is δ_0. This gives rise to a map $x^*: U \to \mathbb{C}^{n+1} - \{0\}$ such that

$$x^*(z) = (y_0(z), z^{\delta_1 - \delta_0} y_1(z), \ldots, z^{\delta_n - \delta_0} y_n(z)).$$

We now claim that $\pi \circ x^*: U \to P_n\mathbb{C}$ extends $x: U - \{p\} \to P_n\mathbb{C}$ to all of U. We need only prove two things: (i) $\pi \circ x^* \mid U - \{p\}$ and (ii) $\pi \circ x^*$ is holomorphic. We will do both simultaneously. Let $U_0 = \{[z_0, \ldots, z_n]: z_0 \neq 0\}$ and let $\zeta: U_0 \to \mathbb{C}^n$ be the coordinate function such that $\zeta([z_0, \ldots, z_n]) = (z_1/z_0, \ldots, z_n/z_0)$. To prove (i), it suffices to prove $(\zeta \circ \pi \circ x^*)(z) = (\zeta \circ x)(z)$, for all $z \in U - \{p\}$. But

$$\begin{aligned}(\zeta \circ \pi \circ x^*)(z) &= (\zeta \circ \pi)(y_0(z), z^{\delta_1 - \delta_0} y_1(z), \ldots, z^{\delta_n - \delta_0} y_n(z)) \\ &= \left(z^{\delta_1 - \delta_0} \frac{y_1(z)}{y_0(z)}, \ldots, z^{\delta_n - \delta_0} \frac{y_n(z)}{y_0(z)}\right). \\ &= (\zeta \circ x)(z).\end{aligned}$$

To prove (ii), we must show that $\zeta \circ \pi \circ x^*$ is holomorphic in U. But this is obvious because in the above expression, $y_0, \ldots, y_n$ are all holomorphic functions and y_0 is never zero in U. This proves that we have extended x over p. Since p is an arbitrary point of $V - V'$, we have in fact extended $x: V' \to P_n\mathbb{C}$ to all of V. The extension is obviously unique. Hence,

<u>Lemma 3.1</u>. A holomorphic mapping $\tilde{x}: V \to \mathbb{C}^{n+1}$ which is not identically zero induces a unique holomorphic map $x: V \to P_n\mathbb{C}$.

In general, let $x: V \to P_n\mathbb{C}$ be holomorphic, and let U be an open set of V. A map $x^*: U \to \mathbb{C}^{n+1} - \{0\}$ such that $\pi \circ x^* = x$ on U is called a <u>reduced representation of</u> x <u>in</u> U. In the course of proving Lemma 3.1, we have clearly proved that if $\tilde{x}: U \to \mathbb{C}^{n+1}$ induces $x: U \to P_n\mathbb{C}$, then x admits a reduced representation in U, provided U is sufficiently small. For a later purpose, we want to prove the existence of reduced representations in greater generality and in a simpler way. So let us drop the assumptions that V is open and that $\tilde{x}: U \to \mathbb{C}^{n+1}$ induces x and simply let $x: V \to P_n\mathbb{C}$ be a holomorphic mapping, V arbitrary. Let U be a small open set in V. For definiteness, let us say U is so small that $x(U) \subseteq U_0$, where $U_0 \subseteq P_n\mathbb{C}$ is the open set $U_0 = \{[z_0,\ldots,z_n]: z_0 \neq 0\}$ on which is the coordinate map $\zeta: U_0 \to \mathbb{C}^n$ such that $\zeta([z_0,\ldots,z_n]) = (z_1/z_0,\ldots,z_n/z_0)$. Relative to ζ, x restricted to U decomposes into n holomorphic functions $\zeta \circ x = (x_1,\ldots,x_n)$. Then the mapping $x^*: U \to \mathbb{C}^{n+1} - \{0\}$ such that $x^*(p) = (1,x_1(p),\ldots,x_n(p))$ is clearly holomorphic and satisfies $\pi \circ x^* = x$ on U. In other words, x^* is a reduced representation of x in U. Thus,

<u>Lemma 3.2</u>. Let $x: V \to P_n\mathbb{C}$ be a holomorphic mapping, where V is an arbitrary Riemann surface. If U is a sufficiently small open set in V, then x admits a reduced representation in U.

We now prove the converse of Lemma 3.1.

<u>Lemma 3.3</u>. Given a holomorphic curve $x: V \to P_n\mathbb{C}$,

where V is open, there exists a map $\tilde{x}: V \to \mathbb{C}^{n+1}$ which induces x.

**Proof*. Let $U_0 = \{[z_0,\ldots,z_n]: z_0 \neq 0\}$ and let $\zeta: U_0 \to \mathbb{C}^n$ be the usual bijection: $\zeta([z_0,\ldots,z_n]) = ({}^{z_1}/_{z_0},\ldots,{}^{z_n}/_{z_0})$. It is quite clear that $x(V)$ cannot be contained in every hyperplane of $P_n\mathbb{C}$, so we may as well assume that $x(V)$ is not contained in $z_0 = 0$. In that case, the multi-valued holomorphic function on V: $p \to \langle[1,0,\ldots,0],x(p)\rangle$ has only a discrete set of zeroes. Let V^* be the complement of these zeroes. Then $x: V^* \to U_0 \subseteq P_n\mathbb{C}$ is well-defined and $\zeta \circ x: V^* \to \mathbb{C}^n$ is then a collection of n holomorphic functions on V^*, say, $\zeta \circ x = (f_1,\ldots,f_n)$. We claim that $f_1,\ldots,f_n$ can be extended to be n meromorphic functions on V. To prove this, let $p \in V - V^*$ and let U be an open set containing p so that $U - \{p\} \subseteq V^*$. Now $x(p) = [0,a_1,\ldots,a_n]$, and there is at least one $a_j \neq 0$. Consider $U_j = \{[z_0,\ldots,z_n]: z_j \neq 0\}$. Then by continuity, x maps some neighborhood of p into U_j. Shrink U if necessary, let $x: U \to U_j$ be well-defined. On U_j is the coordinate function $\xi: U_j \to \mathbb{C}^n$ such that $\xi([z_0,\ldots,z_n]) = (\frac{z_0}{z_j},\ldots,\frac{z_{j-1}}{z_j},\frac{z_{j+1}}{z_j},\ldots,\frac{z_n}{z_j})$, then $\xi \circ x: U \to \mathbb{C}^n$ gives rise to n holomorphic functions on U: $\xi \circ x = (h_0,\ldots,h_{j-1},h_{j+1},\ldots,h_n)$. If we denote by $x^*: U \to \mathbb{C}^{n+1} - \{0\}$ the map $x^*(q) = (h_0(q),\ldots,h_{j-1}(q),1,h_{j+1}(q),\ldots,h_n(q))$ for $q \in U$, then $\pi \circ x^* = x$; for clearly $\xi \circ \pi \circ x^* = \xi \circ x$ and ξ is a bijection. Since $h_0(p) = 0$, we may assume that p is the only zero of h_0 in U (again shrink U if necessary). It follows that for $q \in U - \{p\}$:

$$\begin{aligned}\cdot\ (\zeta \circ x)(q) &= (\zeta \circ \pi \circ x^*)(q) \\ &= \left(\frac{h_1(q)}{h_0(q)}, \ldots, \frac{h_{j-1}(q)}{h_0(q)}, \frac{1}{h_0(q)}, \frac{h_{j+1}(q)}{h_0(q)}, \ldots, \frac{h_n(q)}{h_0(q)}\right).\end{aligned}$$

But $(\zeta \circ x)(q) = (f_1(q), \ldots, f_n(q))$ by definition of $f_1, \ldots, f_n$, so the two ordered sequences of holomorphic functions on $U - \{p\}$, $(f_1, \ldots, f_n)$ and $\left(\frac{h_1}{h_0}, \ldots, \frac{h_{j-1}}{h_0}, \frac{1}{h_0}, \frac{h_{j+1}}{h_0}, \ldots, \frac{h_n}{h_0}\right)$, are in fact equal. As the latter are in fact meromorphic functions on all of U, they are therefore the unique extension to U of $(f_1, \ldots, f_n)$ on $U - \{p\}$. Do this for every $p \in V - V^*$ and we see that $\zeta \circ x = (f_1, \ldots, f_n)$ is extended uniquely to n meromorphic functions $\widetilde{\zeta \circ x} = (\tilde{f}_1, \ldots, \tilde{f}_n)$ on V, i.e., $\widetilde{\zeta \circ x}\colon V \to \mathbb{C} \cup \{\infty\} \times \cdots \times \mathbb{C} \cup \{\infty\}$ is meromorphic.

By the theorem of Behnke-Stein, there exist holomorphic functions $\{\alpha_1, \beta_1, \ldots, \alpha_n, \beta_n\}$ such that $\tilde{f}_i = (\alpha_i/\beta_i)$ on V. Now define:

$$\begin{cases} x_0 = \beta_1 \cdots \beta_n \\ x_i = \beta_1 \cdots \beta_{i-1} \alpha_i \beta_{i+1} \cdots \beta_n, \qquad i = 1, \ldots, n, \end{cases}$$

and let $\tilde{x}\colon V \to \mathbb{C}^{n+1}$ be the holomorphic map $\tilde{x} = (x_0, \ldots, x_n)$. The claim now is that $\tilde{x}$ induces x.

It is perfectly simple to prove this. Let V' be the complement of the common zeroes of $x_0, \ldots, x_n$ in V and recall that V^* is the complement of the discrete set $x^{-1}(P_n\mathbb{C} - U_0)$. Then of course $\pi \circ \tilde{x}\colon V^* \cap V' \to U_0$ and $x\colon V^* \cap V' \to U_0$. We need only prove that $\pi \circ \tilde{x}$ and x agree on $V^* \cap V'$ because the latter is open in V. Now for $p \in V^* \cap V'$,

$$\begin{aligned}(\zeta \circ \pi \circ \tilde{x})(p) &= (\zeta \circ \pi)(x_0(p), \dots, x_n(p)) \\ &= \left(\frac{\alpha_1(p)}{\beta_1(p)}, \dots, \frac{\alpha_n(p)}{\beta_n(p)}\right) = (\tilde{f}_1(p), \dots, \tilde{f}_n(p)) \\ &= (f_1(p), \dots, f_n(p)) \quad \text{because} \quad p \in V^* \\ &= (\zeta \circ x)(p)\end{aligned}$$

by definition of $f_1, \dots, f_n$. Since ζ is a homeomorphism, $\pi \circ \tilde{x}(p) = x(p)$ for all $p \in V^* \cap V'$. Q.E.D.

Because of Lemma 3.3, we will always assume that our given $x: V \to P_n\mathbb{C}$ is induced by a certain $\tilde{x}: V \to \mathbb{C}^{n+1}$. We will eventually need both x and $\tilde{x}$ in most situations, so we should keep (3.1) in mind. Note that $\tilde{x}$ is far from being uniquely determined by x. In fact, it is not difficult to see that $\tilde{x}: V \to \mathbb{C}^{n+1}$ and $\tilde{y}: V \to \mathbb{C}^{n+1}$ induce the same map $V \to P_n\mathbb{C}$ iff there exist nonconstant holomorphic functions h and g on V such that $h\tilde{x} = g\tilde{y}$. To avoid non-essential complications, however, we will agree that <u>once one choice of</u> $\tilde{x}$ <u>is made, it will be kept fixed throughout the discussion.</u>

§2. For the most part, we shall be interested only in those $x: V \to P_n\mathbb{C}$ such that $x(V)$ does not lie in lower dimensional projective subspaces of $P_n\mathbb{C}$. This is a natural condition to impose since we are interested in the question of how many hyperplanes $x(V)$ can fail to meet. If the only subspace of $P_n\mathbb{C}$ that contains $x(V)$ is $P_n\mathbb{C}$ itself, then we say $x: V \to P_n\mathbb{C}$ is <u>nondegenerate</u>; otherwise, <u>degenerate</u>. Clearly x <u>is nondegenerate iff</u> $\tilde{x}: V \to \mathbb{C}^{n+1}$ <u>is not contained</u>

in a proper subspace of $\mathbb{C}^{n+1}$. We need an analytic criterion for nondegeneracy. To this end and for other purposes, we introduce the following quantities: fix a coordinate neighborhood U in V and a coordinate function z on U, we let

$$(3.2)\qquad \begin{cases} X_z^0 = \tilde{x} \\ X_z^k = \tilde{x}\wedge\tilde{x}^{(1)}\wedge\cdots\wedge\tilde{x}^{(k)}, \qquad k = 1,\ldots,n \\ \qquad \text{where by definition } \tilde{x}^{(j)} = \left(\dfrac{d^j x_0}{dz^j},\ldots,\dfrac{d^j x_n}{dz^j}\right) \end{cases}$$

We record here the effect on X_z^k of a different choice of coordinate function on U. Assume a second coordinate function w is given on U, then it is clear that $(\frac{dz}{dw})$ is nowhere zero on U. If we let

$$y^{(k)} = \left(\frac{d^k x_0}{dw^k},\ldots,\frac{d^k x_n}{dw^k}\right), \qquad k = 1,\ldots,n,$$

then obviously,

$$\tilde{y}^{(1)} = \frac{dz}{dw}\,\tilde{x}^{(1)},$$

$$\tilde{y}^{(2)} = \frac{d}{dw}\,\tilde{y}^{(1)} = \left(\frac{dz}{dw}\right)^2\tilde{x}^{(2)} + \frac{d^2z}{dw^2}\,\tilde{x}^{(1)},$$

and in general,

$$\tilde{y}^{(k)} = \left(\frac{dz}{dw}\right)^k\tilde{x}^{(k)} + (\text{terms involving } \tilde{x},\tilde{x}^{(1)},\ldots,\tilde{x}^{(k-1)}).$$

If in analogy with above, we introduce the notation

$$X_w^k = \tilde{y}\wedge\tilde{y}^{(1)}\wedge\cdots\wedge\tilde{y}^{(k)},$$

then

(3.3) $$X_w^k = \left(\frac{dz}{dw}\right)^{\frac{k(k+1)}{2}} X_z^k.$$

We formalize here the convention that the subscript z in the symbol X_z^k will always indicate the local coordinate function with respect to which X_z^k is formed, as in (3.2); when the particular coordinate function used is immaterial, we will omit the subscript and simply write X^k.

Now fix a coordinate neighborhood U and a coordinate function z on U. Then the function on U: $p \mapsto X_z^k(p)$ takes value in $\mathbb{C}^{\ell(k)} \cong \Lambda^{k+1}\mathbb{C}^{n+1}$ (where as usual $\ell(k) = \binom{n+1}{k+1}$, see §2 of Chapter I). If a second coordinate function w is used on U, then X_w^k is related to X_z^k by (3.3). Since $(\frac{dz}{dw})$ is nowhere zero in U, we have proved

Lemma 3.4. The holomorphic mapping $U \to \mathbb{C}^{\ell(k)}$ given by $p \mapsto X^k(p)$ is defined up to multiplication by a zero-free holomorphic function on U.

As a consequence of this lemma, it makes perfect sense to speak of the zeroes of X^k on V. In particular, we can say: "X^n is identically zero on V". (Notation: $X^n \equiv 0$). The criterion of nondegeneracy which we are after is then the following:

Lemma 3.5. $x: V \to P_n\mathbb{C}$ is degenerate iff $X^n \equiv 0$.

Proof. Suppose x is degenerate, then $\tilde{x}$ lies in a subspace of dimension n in $\mathbb{C}^{n+1}$. Over any coordinate neighborhood U of V with coordinate function z, $\tilde{x}^{(k)}$ can be defined as in (3.2) and $\tilde{x}^{(k)}$ lies in the same subspace of

$\mathbb{C}^{n+1}$ as $\tilde{x}$ for $k = 1,\ldots,n$. Since there are $n + 1$ vectors in $\tilde{x},\tilde{x}^{(1)},\ldots,\tilde{x}^{(n)}$, $X_z^n = \tilde{x}\wedge\tilde{x}^{(1)}\wedge\ldots\wedge\tilde{x}^{(n)}$ has to be zero over U and hence $X_z^n \equiv 0$ in U. Consequently, $X^n \equiv 0$ in V. Conversely, suppose $X^n \equiv 0$. As above, let us operate in a coordinate neighborhood U with coordinate function z so that $\tilde{x}^{(k)}$ $(k = 1,\ldots,n)$ are defined as in (3.2). By (1.8),

$$X_z^n = \det\begin{bmatrix} x_0 \cdots\cdots x_n \\ x_0^{(1)}\cdots\cdots x_n^{(1)} \\ \vdots \qquad\qquad \vdots \\ x_0^{(n)}\cdots\cdots x_n^{(n)} \end{bmatrix} \epsilon_0\wedge\ldots\wedge\epsilon_n$$

where $\{\epsilon_0,\ldots,\epsilon_n\}$ is the canonical basis of $\mathbb{C}^{n+1}$ and $x_j^{(i)} = \dfrac{d^i x_j}{dz^i}$. Since $X_z^n \equiv 0$, there is an integer ℓ, $0 \leq \ell < n$, such that

$$A_\ell \equiv \det\begin{bmatrix} x_0 \cdots x_\ell \\ \vdots \qquad \vdots \\ x_0^{(\ell)}\cdots\tilde{x}_\ell^{(\ell)} \end{bmatrix} \not\equiv 0$$

while

$$A_{\ell+1} \equiv \det\begin{bmatrix} x_0 \cdots x_{\ell+1} \\ \vdots \qquad \vdots \\ x_0^{(\ell+1)}\cdots x_{\ell+1}^{(\ell+1)} \end{bmatrix} \equiv 0.$$

Shrink U if necessary, we will assume that A_ℓ is nowhere zero in U. Expand $A_{\ell+1}$ by the last column; this gives a differential equation, homogeneous of order $(\ell+1)$, which is satisfied by $x_{\ell+1}$.

Because of the alternating property of determinants, this same differential equation is also satisfied by $x_0, \ldots, x_\ell$. Hence, $x_0, \ldots, x_{\ell+1}$ are solutions of a homogeneous differential equation of order $(\ell+1)$ whose Wronskian $(= A_{\ell+1})$ vanishes. By a basic theorem of such equations, $x_0, \ldots, x_{\ell+1}$ are linearly dependent in U. Since U is an open set, $x_0, \ldots, x_{\ell+1}$ are linearly dependent in V (because these are holomorphic functions.) Consequently $\tilde{x}(V)$ has to lie in a proper subspace of $\mathbb{C}^{n+1}$, and x is degenerate. Q.E.D.

§3. <u>From now on, we will assume that our holomorphic curve</u> $x\colon V \to P_n\mathbb{C}$ <u>is nondegenerate.</u> To every nondegenerate holomorphic curve is associated a whole series of holomorphic mappings into various affiliated projective spaces of $P_n\mathbb{C}$, and we now describe some of these.

The first such is the so-called <u>associated holomorphic curve of rank</u> k, ${}_kx\colon V \to G(n,k) \subseteq P_{\ell(k)-1}\mathbb{C}$, $k = 0, \ldots, n$. $\left(\ell(k) = \binom{n+1}{k+1}\right)$. Let U be a coordinate neighborhood with coordinate function z, then $X_z^k = \tilde{x} \wedge \tilde{x}^{(1)} \wedge \ldots \wedge \tilde{x}^{(k)}$ can be defined as in (3.2). Identifying $\Lambda^{k+1}\mathbb{C}^{n+1}$ with $\mathbb{C}^{\ell(k)}$ as usual, we have defined a map $X_z^k\colon U \to \mathbb{C}^{\ell(k)}$. By Lemma 3.5, X_z^k is not identically zero. X_z^k therefore induces a map ${}_kx\colon U \to P_{\ell(k)-1}\mathbb{C}$, by Lemma 3.1. Suppose U' is a second coordinate neighborhood with coordinate function z', then the above procedure leads to a map ${}_kx'\colon U' \to P_{\ell(k)-1}\mathbb{C}$. The content of Lemma 3.4 is clearly that ${}_kx$ and ${}_kx'$ agree on $U \cap U'$. Hence the collection of local maps $X_z^k\colon U \to \mathbb{C}^{\ell(k)}$ in fact defines a unique holomorphic mapping ${}_kx\colon V \to P_{\ell(k)-1}\mathbb{C}$. We can say more: since each X_z^k is a decomposable $(k+1)$-vector, ${}_kx(V)$ in fact lies in $G(n,k)$. (See §2 of Chapter I; recall that we identify $G(n,k)$ with its image in $P_{\ell(k)-1}\mathbb{C}$.) The holomorphic mapping ${}_kx\colon V \to G(n,k) \subseteq P_{\ell(k)-1}\mathbb{C}$, $k = 0,\ldots,n$, is called the <u>associated holomorphic curve of rank</u> k. So by the definition of ${}_kx$, if U is any coordinate neighborhood with any coordinate function z, X_z^k defined as in (3.2) and U' is the complement of the zeroes of X_z^k in U (which makes sense by Lemma 3.4), then

$$\begin{array}{ccc} & & \mathbb{C}^{\ell(k)}-\{0\} \\ & \overset{X_z^k}{\nearrow} & \big\downarrow {}_k\pi \\ U' & \xrightarrow{\;{}_kx\;} & G(n,k) \subseteq P_{\ell(k)-1}\mathbb{C} \end{array} \tag{3.4}$$

is a commutative diagram.

Note that in the above, if $k = n$, the whole situation is trivial. So we only get n nontrivial curves: $x \equiv {}_0x, {}_1x, \ldots, {}_{n-1}x$. <u>From now on, we exclude the case</u> $k = n$ <u>from our consideration.</u>

The associated curve of rank k of the nondegenerate curve $x\colon V \to P_n\mathbb{C}$ may well be degenerate, i.e., ${}_kx(V)$ may lie in a lower dimensional projective subspace of $P_{\ell(k)-1}\mathbb{C}$. However, we now show that ${}_kx(V)$ can never lie in a polar divisor Σ_B of $G(n,k)$ if x is nondegenerate. Recall from Chapter I §2 that for $B \in G(n,k)$,

$$(3.5)\qquad \Sigma_B = \{\Lambda \in G(n,k) : \Lambda \cap B^{\perp} \neq \emptyset\}$$
$$= \text{intersection in } P_{\ell(k)-1}\mathbb{C} \text{ of } G(n,k) \text{ and the hyperplane defined by } \langle M,B\rangle = 0 \; (M \in \Lambda^{k+1}\mathbb{C}^{n+1}).$$

So suppose for one $B \in G(n,k)$, ${}_kx(V) \subseteq \Sigma_B$. Equivalently, $\langle {}_kx(V),B\rangle = 0$. Let U be a coordinate neighborhood on which is a coordinate function z such that X_z^k is defined as in (3.2). Shrink U if necessary, we may assume X_z^k has no zero in U. By (3.4), the map $X_z^k\colon U \to \mathbb{C}^{\ell(k)} - \{0\}$ has the property that ${}_k\pi \circ X_z^k = {}_kx$. Thus $X_z^k(p)$ is a representative of ${}_kx(p)$ for each $p \in U$ and hence by the above, $\langle X_z^k,B\rangle \equiv 0$ on U. Let $e_0,\ldots,e_n$ be an O.N. basis of $\mathbb{C}^{n+1}$ such that $\{e_0,\ldots,e_k\}$ spans B. Since $\langle X_z^k,B\rangle = \langle \tilde{x}\wedge\tilde{x}^{(1)}\wedge\cdots\wedge\tilde{x}^{(k)}, e_0\wedge\cdots\wedge e_k\rangle$, (1.9) implies that

$$\det\begin{bmatrix} \langle\tilde{x},e_0\rangle \cdots\cdots \langle\tilde{x},e_k\rangle \\ \vdots \qquad\qquad \vdots \\ \langle\tilde{x}^{(k)},e_0\rangle \cdots \langle\tilde{x}^{(k)},e_k\rangle \end{bmatrix} \equiv 0$$

on U. Let $\tilde{x} = \Sigma_{A=0}^n\, y_A e_A$, then this is equivalent to

$$\det\begin{bmatrix} y_0 \cdots\cdots y_k \\ y_0^{(1)} \cdots y_k^{(1)} \\ \vdots \qquad \vdots \\ y_0^{(k)} \cdots y_k^{(k)} \end{bmatrix} \equiv 0$$

where $y_j^{(i)} = \dfrac{d^i y_j}{dz^i}$. By Lemma 3.5 (or rather, by its proof), $y_0,\ldots,y_k$ are linearly dependent in U. A fortiori,

$y_0,\dots,y_n$ are linearly dependent in U, and hence in all of V. Thus $\tilde{x}(V)$ lies in a proper subspace of $\mathbb{C}^{n+1}$, contradicting the nondegeneracy of x. Hence we have proved

Lemma 3.6. If $x\colon V \to P_n\mathbb{C}$ is nondegenerate, the associated holomorphic curve of rank k $(k = 0,\dots,n-1)$ does not lie in a polar divisor Σ_B for any $B \in G(n,k)$.

§4. A second type of related curve of $x\colon V \to P_n\mathbb{C}$ is obtained by projecting x into an h-dimensional projective subspace A^h of $P_n\mathbb{C}$. More precisely, let $\tilde{x}\colon V \to \mathbb{C}^{n+1}$ be the fixed holomorphic map which induces x, and let $\{e_0,\dots,e_n\}$ be an O.N. basis of $\mathbb{C}^{n+1}$ such that $A^h = e_0 \wedge \cdots \wedge e_h$. Let E stand for the subspace of $\mathbb{C}^{n+1}$ spanned by $e_0,\dots,e_h$. If $\tilde{x} = \Sigma_{A=0}^{n} y_A e_A$, then the projection curve of x into A^h is, by definition, the holomorphic mapping ${}_A x\colon V \to A$ induced by the map ${}_A\tilde{x}\colon V \to E$, where ${}_A\tilde{x}(p) = \Sigma_{a=0}^{h} y_a(p)e_a$. Obviously, if x is nondegenerate, then so is ${}_A x$.

Lemma 3.7. Let U be a coordinate neighborhood with coordinate function z, so that X_z^k and ${}_A X_z^k = {}_A\tilde{x} \wedge {}_A\tilde{x}^{(1)} \wedge \cdots \wedge {}_A\tilde{x}^{(k)}$ are both defined as in (3.2). Then for $k \le h$, $|{}_A X_z^k| = |X_z^k \lrcorner A^h|$. ($\lrcorner$ is the interior product of Chapter 1, §3).

*Proof. Since $\tilde{x} = \Sigma_{A=0}^{n} y_A e_A$, $\tilde{x}^{(j)} = \Sigma_A y_A^{(j)} e_A$, where $y_A^{(j)} = \dfrac{d^j y_A}{dz^j}$. Similarly ${}_A\tilde{x} = \Sigma_{a=0}^{h} y_a e_a$, and ${}_A\tilde{x}^{(j)} = \Sigma_a y_a^{(j)} e_a$. If we expand both ${}_A X_z^k$ and X_z^k in terms of $e_0,\dots,e_n$, it is clear that their coefficients of $e_{i_0} \wedge \cdots \wedge e_{i_k}$ are equal

provided $i_0 < \cdots < i_k \leq h$. $(X_z^k - {}_AX_z^k)$ therefore involves only terms $e_{j_0} \wedge \cdots \wedge e_{j_k}$ where at least one of $j_0, \ldots, j_k$ is bigger than h. But since $A^h = e_0 \wedge \cdots \wedge e_h$, taking the interior product of $(X_z^k - {}_AX_z^k)$ with A^h obviously has the effect of annihilating every single term of the latter type. Thus $(X_z^k - {}_AX_z^k) \lrcorner A^h = 0$, or $X_z^k \lrcorner A^h = {}_AX_z^k \lrcorner A^h$. But clearly $|{}_AX_z^k \lrcorner A^h| = |{}_AX_z^k|$ because ${}_Ax$ lies in A^h.

Q.E.D.

§5. We now come to <u>contracted curve of the first kind.</u> Let A^h be an h-dimensional projective subspace of $P_n\mathbb{C}$ and let an O.N. basis $\{e_0, \ldots, e_n\}$ of $\mathbb{C}^{n+1}$ be chosen so that $A^h = e_0 \wedge \cdots \wedge e_h$. Let U be a coordinate neighborhood on which is a coordinate function z, so that X_z^k is defined as in (3.2). For $h < k \leq n-1$, the function $p \mapsto (X_z^k \lrcorner A^h)(p)$ is a mapping $U \to \Lambda^{k-h}\mathbb{C}^{n+1}$. By Lemma 1.4, $X_z^k \lrcorner A^h$ is a decomposable $(k-h)$ vector and so this induces a holomorphic mapping $U \to G(n,k-h-1) \subseteq P_{\ell(k-h-1)-1}\mathbb{C}$, where $\ell(k-h-1) = \binom{n+1}{k-h}$. (Lemma 3.1). If a different representative $e_0' \wedge \cdots \wedge e_h'$ were used for A^h, and a different coordinate function w were chosen in U, the resulting $X_w^k \lrcorner A^h$ would differ from the $X_z^k \lrcorner A^h$ of above by a nowhere zero holomorphic function (Lemma 3.4). Hence the resulting mapping $U \to G(n,k-h-1 \subseteq P_{\ell(k-h-1)-1}\mathbb{C}$ would have been the same. This implies that we actually have a holomorphic map uniquely defined on all of V: for $h < k \leq n-1$,

$$X^k \lrcorner A^h \colon V \to G(n,k-h-1) \subseteq P_{\ell(k-h-1)-1}\mathbb{C},$$

and we call this a <u>contracted curve of the first kind</u>.

<u>Lemma 3.8.</u> $X^k \lrcorner A^h$ does not lie in a polar divisor of $G(n,k-h-1)$.

<u>Proof.</u> Suppose it lies in Σ_B, where $B \in G(n,k-h-1)$. By (3.5), $\langle X^k \lrcorner A^h, B\rangle = 0$, i.e., $\langle X^k, A^h \wedge B\rangle = 0$. But this means that X^k lies in the polar divisor of $G(n,k)$ defined by the decomposable k-vector $A^h \wedge B$, contradicting Lemma 3.6.

Q.E.D.

The most interesting special case is when $k = h + 1$. In this case, $X^{h+1} \lrcorner A^h : V \to P_n\mathbb{C}$. We denote as usual the polar space of A^h by $A^\perp$ and claim that in fact,

$$X^{h+1} \lrcorner A^h : V \to A^\perp \tag{3.6}$$

This is quite clear because if we let $\{e_o,\ldots,e_n\}$ be an O.N. basis of $\mathbb{C}^{n+1}$ such that $A^h = e_o \wedge \cdots \wedge e_h$, then $A^\perp = e_{h+1} \wedge \cdots \wedge$ By definition of the interior product, if $a = 0,\ldots,h$, then

$$\langle e_\alpha, X^{h+1} \lrcorner A^h\rangle = \langle X^{h+1}, e_o \wedge \cdots \wedge e_h \wedge e_a\rangle = 0.$$

Hence $(X^{h+1} \lrcorner A^h)(V) \subseteq A^\perp$.

Return now to the coordinate neighborhood U with coordinate function z, then $X_z^h, X_z^{h+1}, X_z^{h+2}$ are defined as in (3.2). Let us write $y = X_z^{h+1} \lrcorner A^h$ and $y^{(1)} = \frac{dy}{dz}$. We have the following important lemma.

<u>Lemma 3.9.</u> $y \wedge y^{(1)} = \langle A^h, X_z^h\rangle (A^h \lrcorner X_z^{h+2})$

<u>Proof.</u> We have $X_z^{h+1} = \tilde{x} \wedge \tilde{x}^{(1)} \wedge \cdots \wedge \tilde{x}^{(h)} \wedge \tilde{x}^{(h+1)}$, hence

$$\frac{d}{dz}(X_z^{h+1}) = \tilde{x}\wedge\tilde{x}^{(1)}\wedge\cdots\wedge\tilde{x}^{(h)}\wedge\tilde{x}^{(h+2)}.$$

Consequently,

$$y = (X_z^h\wedge\tilde{x}^{(h+1)}) \lrcorner A^h$$
$$y^{(1)} = (X_z^h\wedge\tilde{x}^{(h+2)}) \lrcorner A^h$$

To prove the lemma, it suffices to show that for any two vectors e and f of $\mathbb{C}^{n+1}$:

$$\langle y\wedge y^{(1)}, e\wedge f\rangle = \langle A^h, X_z^h\rangle\langle X_z^{h+2} \lrcorner A^h, e\wedge f\rangle,$$

i.e.,

$$\langle y,e\rangle\langle y^{(1)},f\rangle - \langle y,f\rangle\langle y^{(1)},e\rangle$$
$$= \langle A^h, X_z^h\rangle\langle A^h\wedge e\wedge f, X_z^{h+2}\rangle.$$

Using the preceding observations, this is equivalent to:

$$(3.7)\qquad \langle X_z^h, A^h\rangle\langle X_z^h\wedge\tilde{x}^{(h+1)}\wedge\tilde{x}^{(h+2)}, A^h\wedge e\wedge f\rangle$$
$$= \langle X_z^h\wedge\tilde{x}^{(h+1)}, A^h\wedge e\rangle\langle X_z^h\wedge\tilde{x}^{(h+2)}, A^h\wedge f\rangle$$
$$- \langle X_z^h\wedge\tilde{x}^{(h+1)}, A^h\wedge f\rangle\langle X_z^h\wedge\tilde{x}^{(h+2)}, A^h\wedge e\rangle.$$

We will prove (3.7) by proving the following general statement: let Θ and Ξ be two decomposable vectors in $\Lambda^p\mathbb{C}^{n+1}$, and let μ_1, μ_2 and ν_1, ν_2 be any four vectors. Then

$$(3.8)\qquad \langle\Theta,\Xi\rangle\langle\Theta\wedge\mu_1\wedge\mu_2, \Xi\wedge\nu_1\wedge\nu_2\rangle$$
$$= \langle\Theta\wedge\mu_1, \Xi\wedge\nu_1\rangle\cdot\langle\Theta\wedge\mu_2, \Xi\wedge\nu_2\rangle$$
$$- \langle\Theta\wedge\mu_1, \Xi\wedge\nu_2\rangle\cdot\langle\Theta\wedge\mu_2, \Xi\wedge\nu_1\rangle$$

Because of definition (1.9) of the inner product for decomposable vectors, (3.8) is equivalent to an identity on determinants, which we formulate as follows. Let A be a $(p+2) \times (p+2)$ matrix, $A = (a_{ij})$, and let A' denote the $p \times p$ submatrix in the upper-left corner of A,

$$\text{i.e., } A' = \begin{bmatrix} a_{11} & \cdots & a_{1p} \\ \vdots & & \vdots \\ a_{p1} & \cdots & a_{pp} \end{bmatrix}.$$

Furthermore, let $A_{\alpha\beta}$ denote the $(p+1) \times (p+1)$ submatrix of A obtained by deleting the α-th row and β-th column. Then

$$(3.9)_{p+2} \qquad \det A \cdot \det A' = \det \begin{bmatrix} \det A_{p+1,p+1} & \det A_{p+1,p+2} \\ \det A_{p+2,p+1} & \det A_{p+2,p+2} \end{bmatrix}.$$

This is known as Sylvester's theorem on compound determinants. We will prove it by induction on p. For $p = 1,2$, both $(3.9)_3$ and $(3.9)_4$ can be proved by a simple computation. Now we assume $p > 2$, and suppose that $(3.9)_3$, $(3.9)_4$, ..., $(3.9)_{p+1}$ have already been proved. Both sides of $(3.9)_{p+2}$ are linear in the $(p+1)$-th and $(p+2)$-th rows and columns, therefore we need only consider those A's with only one 1 and with the rest of the entries equal to zero in any of the $(p+1)$-th and $(p+2)$-th rows and columns. After disposing of the trivial cases, we find that we only have to consider those A's in which the 1's lie outside of the 2×2 box in the lower right corner. Since permutations of the first p rows and the first p columns leave $(3.9)_{p+2}$ invariant, we are reduced to proving $(3.9)_{p+2}$ only for an A of the following form:

$$A = \left[\begin{array}{ccc|cc} a_{11} & \cdots & a_{1p} & 0 & 0 \\ \vdots & & \vdots & \vdots & \vdots \\ \vdots & & \vdots & 0 & \vdots \\ \vdots & & \vdots & 1 & 0 \\ a_{p1} & \cdots & a_{pp} & 0 & 1 \\ \hline 0\cdots 0 & 1 & 0 & 0 & 0 \\ 0\cdots\cdots & 0 & 1 & 0 & 0 \end{array}\right]$$

The matrix in the upper-left corner of A is of course just A'. Using the notation above, let us denote by $A'_{\alpha\beta}$ the submatrix of A' obtained by deleting the α-th row and β-th column. Furthermore, let A'' be the $(p-2) \times (p-2)$ submatrix in the upper-left corner of A',

$$\text{i.e.,} \quad A'' = \begin{bmatrix} a_{11} & \cdots\cdots & a_{1,p-2} \\ \vdots & & \vdots \\ a_{p-2,1} & \cdots & a_{p-2,p-2} \end{bmatrix}.$$

Then for this particular A, Laplace's expansion of each term of $(3.9)_{p+2}$ reduces $(3.9)_{p+2}$ to

$$\det A' \cdot \det A'' = \det \begin{bmatrix} \det A'_{p-1,p-1} & \det A'_{p-1,p} \\ \det A'_{p,p-1} & \det A'_{p,p} \end{bmatrix}$$

But we recognize this as $(3.9)_p$, which is true by the induction assumption. Hence $(3.9)_{p+2}$ is true. Q.E.D.

§6. Finally, we define <u>contracted curve of the second</u>

kind. The details here are almost identical with those of the preceding section and so will be left out. Let A^h be an h-dimensional projective subspace of $P_n\mathbb{C}$ and let U be a coordinate neighborhood with coordinate function z. For $k < h \leq n$, the function $p \mapsto (X_z^k \lrcorner A^h)(p)$ leads to a mapping $U \to \Lambda^{h-k}\mathbb{C}^{n+1}$. By Lemma 1.4, $A^h \lrcorner X_z^k$ is a decomposable (h-k)-vector and so this induces a holomorphic mapping $U \to G(n,h-k-1) \subseteq P_{\ell(h-k-1)-1}\mathbb{C}$. By Lemma 3.4, this gives rise to a well-defined holomorphic curve:

$$X^k \lrcorner A^h\colon V \to G(n,h-k-1) \subseteq P_{\ell(h-k-1)-1}\mathbb{C}$$

for $k < h \leq n$, and we call this a <u>contracted curve of the second kind</u>.

<u>Lemma 3.10.</u> $X^k \lrcorner A^h$ does not lie in a polar divisor of $G(n,h-k-1)$.

We consider the special case $h = k + 1$. Then $X^k \lrcorner A^{k+1}$: $V \to P_n\mathbb{C}$. Actually the image lies in A^{k+1}, so that

$$X^k \lrcorner A^{k+1}\colon V \to A^{k+1}. \tag{3.11}$$

Staying in U, let $y = X_z^k \lrcorner A^{k+1}$ and $y^{(1)} = \frac{dy}{dz}$. Then Sylvester's theorem on compound determinants (3.9) implies

<u>Lemma 3.11.</u> $y \wedge y^{(1)} = \langle X_z^{k+1}, A^{k+1}\rangle (X_z^{k-1} \lrcorner A^{k+1})$.

Lemmas 3.9 and 3.11 will be of critical importance in Chapter V.

CHAPTER IV

The two main theorems for holomorphic curves

§1. The first goal of this chapter is to establish the First Main Theorem for a holomorphic mapping $f: V \to G(n,k) \subseteq P_{\ell(k)-1}\mathbb{C}$, $k = 0,\ldots,n-1$, where $\ell(k) = \binom{n+1}{k+1}$, $G(n,k)$ is the set of k-dimensional projective subspaces of $P_n\mathbb{C}$, and V is an open Riemann surface admitting a harmonic exhaustion. (Definition 2.1). From Chapter III, §3, we learn that each holomorphic curve $x: V \to P_{n+1}\mathbb{C}$ gives rise to a series of associated curves of rank k, ${}_kx: V \to G(n,k) \subseteq P_{\ell(k)-1}\mathbb{C}$, $k = 0,\ldots,n-1$. Thus we are considering all at once the holomorphic curves $x = {}_0x, {}_1x, \ldots, {}_{n-1}x$. Recall from Chapter I, that $G(n,k)$ has a naturally given Kahler metric, which is the restriction of the F-S metric on $P_{\ell(k)-1}\mathbb{C}$. If on $\mathbb{C}^{\ell(k)}$:

$$(4.1) \qquad \tilde{F} = \frac{1}{\langle\Lambda,\Lambda\rangle^2}\{\langle\Lambda,\Lambda\rangle\langle d\Lambda,d\Lambda\rangle - \langle d\Lambda,\Lambda\rangle\langle\Lambda,d\Lambda\rangle\},$$

then ${}_k\pi^* F = \tilde{F}$, where ${}_k\pi: \mathbb{C}^{\ell(k)} \to P_{\ell(k)-1}\mathbb{C}$ is the usual fibration and $\Lambda = (\ldots,\lambda_{i_0\cdots i_k},\ldots)$ is the coordinate function on $\mathbb{C}^{\ell(k)}$. By (1.6), the associated two-form $\tilde{\omega}$ of $\tilde{F}$ has the property that

$$(4.2) \qquad \tilde{\omega} = \frac{1}{2}\, dd^c \log|\Lambda|.$$

If ω is the Kahler form of F, then ${}_k\pi^*\omega = \tilde{\omega}$. Given $B \in G(n,k)$, we have the polar divisor Σ_B on $G(n,k)$ given by (3.5). In particular, if we denote by Π_B the hyperplane defined by $\{\Lambda: \langle\Lambda,B\rangle = 0\}$, then $\Sigma_B = \Pi_B \cap G(n,k)$. Denote by $\tilde{u}_B$ the function defined on $\mathbb{C}^{\ell(k)} - \{0\} - \{{}_k\pi^{-1}(\Pi_B)\}$ such that:

$$\tilde{u}_B(\Lambda) = \log \frac{|\Lambda|}{|\Lambda,B|} \tag{4.3}$$

where we now write $|\Lambda,B|$ for $|\langle\Lambda,B\rangle|$ for convenience. (Note that although the multi-vector B is only defined up to a multiple of $e^{\sqrt{-1}\theta}$, $|\Lambda,B|$ suffers from no such ambiguity.) Because $\langle\Lambda,B\rangle$ is a holomorphic function of Λ, $dd^c|\Lambda,B|$ $= 2\sqrt{-1}\, d'd''\log|\Lambda,B| = \sqrt{-1}\{d'd''\log\langle\Lambda,B\rangle\overline{\langle\Lambda,B\rangle}\} = \sqrt{-1}\{d'd''\log\langle\Lambda,B\rangle - d''d'\log\overline{\langle\Lambda,B\rangle}\} = 0$. Hence (4.2) and (4.3) imply that

$$\begin{aligned}\frac{1}{2}\, dd^c\tilde{u}_B &= \frac{1}{2}\, dd^c \log \frac{|\Lambda|}{|\Lambda,B|} \\ &= \frac{1}{2}\, dd^c \log |\Lambda| - \frac{1}{2}\, dd^c \log |\Lambda,B| \\ &= \frac{1}{2}\, dd^c \log |\Lambda| \\ &= \tilde{\omega}\end{aligned}$$

in $\mathbb{C}^{\ell(k)} - \{0\} - \{{}_k\pi(\Pi_B)\}$. Moreover, for every $\lambda \in \mathbb{C}^*$, clearly $\tilde{u}_B(\lambda\Lambda) = \tilde{u}_B(\Lambda)$, and so there is a C^∞ function u_B defined on $P_{\ell(k)-1}\mathbb{C} - \Pi_B$ such that ${}_k\pi^* u_B = \tilde{u}_B$. Thus on $P_{\ell(k)-1}\mathbb{C} - \Pi_B$, ${}_k\pi^*(\omega - \frac{1}{2}dd^c u_B) = \tilde{\omega} - \frac{1}{2}dd^c\tilde{u}_B = 0$. Since ${}_k\pi^*$ is injective (because $d_k\pi$ is surjective), $\omega = \frac{1}{2}\, dd^c u_B$ on $P_{\ell(k)-1}\mathbb{C} - \Pi_B$.

Now restrict all this information to $G(n,k)$. If we agree to denote the restriction of F, ω, u_B, etc. to $G(n,k)$ still by the same letters , then we have clearly proved:

Theorem 4.1. If $B \in G(n,k)$, $k = 0,\ldots,n-1$, there is a function u_B such that:

(i) u_B is C^∞ on $G(n,k) - \Sigma_B$.

(ii) $\frac{1}{2} dd^c u_B = \omega_B$ in $G(n,k) - \Sigma_B$, where ω is the Kahler form of the restriction of the F-S metric to $G(n,k)$.

(iii) If we denote by $\tilde{u}_B$ the function ${}_k\pi^* u_B$, where ${}_k\pi\colon \mathbb{C}^{\ell(k)} - \{0\} \to P_{\ell(k)-1}\mathbb{C}$, then $\tilde{u}_B(\Lambda) = \log \frac{|\Lambda|}{|\Lambda, B|}$.

Now we return to the consideration of a holomorphic mapping $f\colon V \to G(n,k) \subseteq P_{\ell(k)-1}\mathbb{C}$. For the moment, we need not assume V to be open. Let D be a compact surface with C^∞ boundary in V and assume that $f(V)$ does not lie in Σ_B for a fixed $B \in G(n,k)$. Thus the multi-valued holomorphic function on D such that $p \mapsto \langle f(p), B\rangle$ is not identically zero and consequently its zeroes are isolated. We define

(4.4) $\quad n(D,B)$ = sum of the orders of zeroes of $\langle f, B\rangle$ in D.

We now give a motivation for this definition. If $x\colon V \to P_n\mathbb{C}$ is our original holomorphic curve, the prime question of interest in equidistribution theory is: given a hyperplane Π of $P_n\mathbb{C}$, does $x(V)$ intersect Π? Let a be the point of $P_n\mathbb{C}$ which is the polar subspace of Π. Then clearly, $x(V)$ intersects Π if and only if $\langle x, a\rangle$ has a zero in V. More generally, let ${}_kx\colon V \to G(n,k) \subseteq P_{\ell(k)-1}\mathbb{C}$ be the associated curve of rank k. For each $p \in V$, ${}_kx(p)$ is a k-dimensional projective subspace of $P_n\mathbb{C}$ and so $\bigcup_{p\in V} {}_kx(p)$ is a subset of $P_n\mathbb{C}$. By abuse of notation, we also denote this union by ${}_kx(V)$. (One can in fact show that ${}_kx(V)$ is locally a (k + 1)-dimensional subvariety of $P_n\mathbb{C}$). So given an (n-k-1)-dimensional projective subspace $B^\perp$ of $P_n\mathbb{C}$, we want to know if ${}_kx(V)$ intersects $B^\perp$ or not. Rephrasing this a little, let B be the polar space of $B^\perp$, then $B \in G(n,k)$. By (3.5), the above question is equivalent to: does the multi-valued holomorphic function $p \mapsto \langle {}_kx(p), B\rangle$ have a zero in V? One can in fact prove that the intersection number of the singular chains $B^\perp$ and ${}_kx(D)$ in $P_n\mathbb{C}$ is exactly equal to the sum of the orders of zeroes of $\langle {}_kx, B\rangle$ in D. Thus we are led to the consideration of $n(D,B)$ as given in (4.4).

Now define $v(D) = \frac{1}{\pi}\int_D f^*\omega$, where ω is the Kahler form of the Fubini-Study metric of $P_{\ell(k)-1}\mathbb{C}$. The factor $\frac{1}{\pi}$ is due to the fact that <u>if</u> $P_1\mathbb{C}$ <u>is any one-dimensional projective subspace of</u> $P_{\ell(k)-1}\mathbb{C}$, <u>then</u> $\int_{P_1\mathbb{C}} \omega = \pi$. This easily proven fact will also follow from §1 of Chapter V. The following theorem is then the non-integrated First Main Theorem.

<u>Theorem 4.2</u>. Let $f: V \to G(n,k) \subseteq P_{\ell(k)-1}\mathbb{C}$, $k = 0,\ldots,n-1$, be holomorphic and V is arbitrary. Assume that for some $B \in G(n,k)$, $f(V)$ does not lie in the polar divisor Σ_B. If D is a compact subdomain with C^∞ boundary in V such that $f(\partial D) \cap \Sigma_B = \emptyset$, then

$$n(D,B) + \int_{\partial D} f^*\lambda_B = v(D), \tag{4.5}$$

where $\lambda_B = \frac{1}{2\pi} d^c u_B$. (See Theorem 4.1).

<u>Proof</u>. Let $g: D \to \mathbb{R}$ be the real-valued function

$g(p) = |f(p),B|$. Since $f(D)$ does not lie in Σ_B, the zeroes of g are isolated in D and hence form a finite set $\{a_1,\dots,a_m\}$. Since $f(\partial D) \cap \Sigma_B = \emptyset$, $\{a_1,\dots,a_m\} \subseteq D - \partial D$. Hence we may choose a neighborhood U_ϵ of zero in $\mathbb{R}$ with the property that $g^{-1}(U_\epsilon) = U_1 \cup \cdots \cup U_m$, where $a_j \in U_j$ and $U_j \cap U_k = \emptyset$ if $j \neq k$. Thus $f(D - U_1 \cup \cdots \cup U_m)$ is disjoint from Σ_B. By Stokes' theorem and Theorem 4.1,

$$\begin{aligned} v(D) &= \frac{1}{\pi}\int_D f^*\omega \\ &= \frac{1}{\pi}\lim_{\epsilon\to 0}\int_{D-U_1-\cdots-U_m} f^*\omega \\ &= \frac{1}{\pi}\lim_{\epsilon\to 0}\int_{D-U_1-\cdots-U_m} f^*\left(\frac{1}{2}\,dd^c u_B\right) \\ &= \lim_{\epsilon\to 0}\int_{D-U_1-\cdots-U_m} df^*\lambda_B \\ &= \int_{\partial D} f^*\lambda_a + \lim_{\epsilon\to 0}\Sigma_{j=1}^m - \int_{\partial U_j} f^*\lambda_B \end{aligned}$$

It remains to prove that the last sum is equal to $n(D,B)$. For this purpose, it is sufficient to prove that

(4.6) $$\text{the order of zero of } \langle f(a_j),B\rangle = \lim_{\epsilon\to 0} - \int_{\partial U_j} f^*\lambda_B .$$

We may clearly assume that U_j is very small so that by Lemma 3.2, there is a reduced representation of f in U_j, i.e., there is a holomorphic map $y\colon U_j \to \mathbb{C}^{\ell(k)} - \{0\}$ such that ${}_k\pi \circ y = f$. Now choose O.N. bases $\{e_0,\dots,e_n\}$ in $\mathbb{C}^{n+1}$ so that $B = e_0\wedge\cdots\wedge e_k$. Write $y = y_1 e_0\wedge\cdots\wedge e_k + \cdots$, then for every $p \in U_j$, $\langle y(p), B\rangle = y_1(p)$. By (iii) of Theorem 4.1

$$f^*u_B = y^*{}_k\pi^* u_B = y^* \log \frac{|\Lambda|}{|\Lambda, B|} = \log \frac{|y|}{|y, B|}$$
$$= \log \frac{|y|}{|y_1|} = \log |y| - \log |y_1|,$$

so that,

$$\int_{\partial U_j} -f^*\lambda_B = \frac{1}{2\pi} \int_{\partial U_j} -d^c f^* u_B$$
$$= \frac{1}{2\pi} \int_{\partial U_j} d^c \log |y_1| - \frac{1}{2\pi} \int_{\partial U_j} d^c \log |y|.$$

Since $|y|$ is never zero, $d^c \log |y|$ is clearly C^∞ in U_j, so that $\lim_{\epsilon\to 0} \int_{\partial U_j} d^c \log |y| = 0$. Furthermore, $y(a_j)$ is a representative of the projective subspace $f(a_j)$ of $P_n\mathbb{C}$, hence $|f(a_j), B| = |y(a_j), B| = |y_1(a_j)|$. So the order of zero of $\langle f(a_j), B\rangle$ is equal to the order of zero of y_1 at a_j. To prove (4.6), it suffices to prove:

(4.7) the order of zero of $y_1(a_j)$

$$= \lim_{\epsilon\to 0} \frac{1}{2\pi} \int_{\partial V_j} d^c \log |y_1|.$$

This is essentially the argument principle. In greater detail, let z be a local coordinate function centered at a_j. Then there is an integer $m(j)$ such that $y_1(z) = z^{m(j)} h(z)$, where h is holomorphic and $h(0) \neq 0$. We may assume U_j is so small that h is nowhere zero in U_j. So the order of zero of $y_1(a_j)$ is just $m(j)$. Now $d^c \log |y_1| = m(j) d^c \log |z| + d^c \log |h| = m(j) d\theta + d^c \log |h|$, where $z = |z| e^{\sqrt{-1}\theta}$.

(See (2.7)). Remembering that $\log |h|$ is C^∞ in U_j because h is nowhere zero, we have

$$\lim_{\epsilon\to 0} \frac{1}{2\pi} \int_{\partial U_j} d^c \log |y_1| = \frac{1}{2\pi} \lim_{\epsilon\to 0} \int_{\partial U_j} m(j)\, d\theta + \lim_{\epsilon\to 0} \frac{1}{2\pi} \int_{\partial U_j} d^c \log |h|$$

$$= \frac{1}{2\pi} m(j) \cdot 2\pi + 0 = m(j).$$

This proves (4.7) and there with the theorem. Q.E.D.

§2. Now we assume V is open and has a harmonic exhaustion function (Definition 2.1) τ which is harmonic on $\{p: p \in V, \tau(p) \geq r(\tau)\}$. We recall this notation: $V[r] = \{p: p \in V, \tau(p) \leq r\}$, $\partial V[r] = \{p: \tau(p) = r\}$. We shall work exclusively in $V - V[r(\tau)]$, i.e., only in the domain of harmonicity of τ, so that <u>all parameter values</u> r <u>are assumed greater than</u> $r(\tau)$. In $V - V[r(\tau)]$, the critical points of τ are isolated. Also recall that if $p \in V - V[r(\tau)]$ and if $d\tau(p) \neq 0$, then in a sufficiently small neighborhood of p, there is a holomorphic function $\sigma = \tau + \sqrt{-1}\rho$ which serves as a coordinate function. (Lemma 2.4 and the remarks after Definition 2.2)

Now return to our previous situation. We have $f: V \to G(n,k) \subseteq P_{\ell(k)-1}\mathbb{C}$ and f is holomorphic. If we assume that for a fixed $B \in G(n,k)$, $f(V)$ does not lie in Σ_B, and furthermore that

($\mathcal{A}$) $f(\partial V[r]) \cap \Sigma_B = \emptyset$,

($\mathcal{B}$) r is not a critical value of τ,

then Theorem 4.2 implies that

$$n(r,B) + \int_{\partial V[r]} f^*\lambda_B = v(r)$$

where we have written $n(r,B)$ for $n(V[r],B)$, $v(r)$ for $v(V[r])$ and $\lambda_B = \frac{1}{2\pi} d^c u_B$. The use of the special coordinate function $\sigma = \tau + \sqrt{-1}\rho$ leads to:

Lemma 4.3. Under assumptions $(\mathfrak{A})$ and $(\mathcal{L})$,

$$\int_{\partial V[r]} f^*\lambda_B = \frac{d}{dr}\Big(\frac{1}{2\pi} \int_{\partial V[r]} f^* u_B * d\tau\Big).$$

The proof is identical with that of Lemma 2.5. Thus we have that if $(\mathfrak{A})$ and $(\mathcal{L})$ hold for r, then

$$n(r,B) + \frac{d}{dr}\Big(\frac{1}{2\pi} \int_{\partial V[r]} f^* u_B * d\tau\Big) = v(r).$$

Suppose now $[r_1,r_2]$ is an interval in which $(\mathfrak{A})$ and $(\mathcal{L})$ hold for all $r \in [r_1,r_2]$, then an integration leads to:

$$(4.8) \qquad \int_{r_1}^{r_2} n(t,B)dt + \frac{1}{2\pi} \int_{\partial V[t]} f^* u_B * d\tau \Big|_{r_1}^{r_2} = \int_{r_1}^{r_2} v(t)dt,$$

where we have used the standard notation: $h(t)\Big|_{r_1}^{r_2} = h(r_2) - h(r_1)$. To extend (4.8) to all subintervals of $(r(\tau),s)$ without the restrictions of $(\mathfrak{A})$ and $(\mathcal{L})$ we need the following

Lemma 4.4. $\int_{\partial V[r]} f^* u_B * d\tau$ for a fixed B is a continuous function of r for all r.

*Proof. We first show that the lemma follows from the following two facts:

(α) $\int_{V[r]} d(f^* u_B * d\tau)$ is finite for all r.

(β) If $f(p) \in \Sigma_B$ and W_ϵ is a holomorphic disc $\{|z| < \epsilon\}$ about p such that $z(p) = 0$, then $\lim_{\epsilon \to 0} \int_{\partial W_\epsilon} f^* u_B * d\tau = 0$.

As always, we invoke Stokes' theorem. So let $f^{-1}(\Sigma_B) \cap V[r] = \{p_1, \ldots, p_\ell\}$. Enclose each p_i by a holomorphic disc $U_\epsilon^i = \{|z_i| < \epsilon\}$, where $z_i(p_i) = 0$, and let $U_\epsilon = \bigcup_{i=1}^{\ell} U_\epsilon^i$. On $V[r] - U_\epsilon$, $f^* u_B$ is C^∞ and so $f^* u_B * d\tau$ is a C^∞ one-form. Thus,

$$\int_{V[r]} d(f^* u_B * d\tau) = \lim_{\epsilon \to 0} \int_{V[r]-U_\epsilon} d(f^* u_B * d\tau) \qquad \text{(by } (\alpha))$$

$$= \lim_{\epsilon \to 0} \{\int_{\partial V_\epsilon[r]} f^* u_B * d\tau - \Sigma_{i=1}^{\ell} \int_{\partial U_\epsilon^i} f^* u_B * d\tau\}$$

(where $\partial V_\epsilon[r]$ denotes the subset of $\partial V[r]$ outside of those U_ϵ^i which intersect $\partial V[r]$; consequently we have $\lim_{\epsilon \to 0} \partial V_\epsilon[r] = \partial V[r]$).

$$= \int_{\partial V[r]} f^* u_B * d\tau \qquad \text{(by } (\beta)).$$

This shows that the last integral is finite for all r. To prove continuity in r, it suffices to show that $r_i \uparrow r$ implies

$$\int_{\partial V[r_i]} f^* u_B * d\tau \to \int_{\partial V[r]} f^* u_B * d\tau. \tag{4.9}$$

If χ_i denotes the characteristic function of $V[r_i]$, then Lebesgue's bounded convergence theorem implies that

$$\begin{aligned}\int_{\partial V[r_1]} f^* u_B * d\tau &= \int_{V[r_1]} d(f^* u_B * d\tau) \\ &= \int_{V[r]} \chi_1 d(f^* u_B * d\tau) \\ &\to \int_{V[r]} d(f^* u_B * d\tau) = \int_{\partial V[r]} f^* u_B * d\tau .\end{aligned}$$

This proves (4.9) and hence the lemma. It remains to prove (α) and (β). The singularities of $d(f^* u_B * d\tau)$ are located at the discrete set $f^{-1}(\Sigma_B)$, and we need only examine this two-form at a $p \in V[r]$ such that $f(p) \in \Sigma_B$. Let U be a coordinate neighborhood at p and let $y\colon U \to \mathbb{C}^{\ell(k)} - \{0\}$ be a reduced representative of f at U. (Lemma 3.2). So $_k\pi \circ y = f$. Choose O.N. basis $\{e_0, \ldots, e_n\}$ in $\mathbb{C}^n$ be chosen so that $B = e_0 \wedge \cdots \wedge e_k$ and write $y = y_1 e_0 \wedge \cdots \wedge e_k + \cdots$. Thus $\langle y, B\rangle = y_1$ and

$$\begin{aligned} f^* u_B = (y^* \circ {}_k\pi^*) u_B = y^* \log \frac{|\Lambda|}{|\Lambda, B|} = \log \frac{|y|}{|y_1|} \\ = \log |y| - \log |y_1| . \end{aligned}$$

Thus $d(f^* u_B * d\tau) = d(\log |y| * d\tau) - d(\log |y_1| * d\tau)$ in U. But $|y|$ is never zero, so $d(\log |y| * d\tau)$ is C^∞ in U. Therefore it suffices to show that $\int_U d(\log |y_1| * d\tau)$ is finite. Let z be a coordinate function centered at p such that $y_1(z) = z^m h(z)$, where h is holomorphic and $h(0) \neq 0$. We may as well assume that h is never zero in U. Then

$$\begin{aligned} d(\log |y_1| * d\tau) &= d\{(m \log |z| + \log |h|) * d\tau\} \\ &= m d(\log |z| * d\tau) + C^\infty \text{ form} \\ &= \frac{m}{|z|} d|z| \wedge * d\tau + m \log |z| d * d\tau + C^\infty \text{ form}. \end{aligned}$$

In terms of polar coordinates, both forms are obviously integrable in U. This proves (α). For (β), we use the same notation and have that in U:

$$f^* u_B {*d\tau} = C^\infty \text{ form} - (\log |y_1| {*d\tau})$$
$$= C^\infty \text{ form} - m \log |z| {*d\tau}.$$

If $\partial W_\epsilon = \{p: |z(p)| = \epsilon\}$, then

$$\int_{\partial W_\epsilon} f^* u_B {*d\tau} = \int_{\partial W_\epsilon} C^\infty \text{ form} - m \log \epsilon \int_{\partial W_\epsilon} {*d\tau}.$$

There is no question that as $\epsilon \to 0$, the first integral on the right approaches zero, so it remains to prove that $\lim_{\epsilon\to 0} \log \epsilon \int_{\partial W_\epsilon} {*d\tau} = 0$. We use polar coordinates $\log r + \sqrt{-1}\theta$ in U minus a radical slit, where $z = re^{\sqrt{-1}\theta}$. Then $*d\tau = *(\frac{\partial\tau}{\partial r} \cdot r d \log r + \frac{\partial\tau}{\partial\theta} d\theta) = r \frac{\partial\tau}{\partial r} d\theta - \frac{\partial\tau}{\partial\theta} d \log r$. Consequently,

$$\lim_{\epsilon\to 0} \log \epsilon \int_{\partial W_\epsilon} {*d\tau} = \lim_{\epsilon\to 0} \log \epsilon \cdot \int_{\partial W_\epsilon} r \frac{\partial\tau}{\partial r} d\theta$$
$$= \lim_{\epsilon\to 0} \epsilon \log \epsilon \int_{\partial W_\epsilon} \frac{\partial\tau}{\partial r} d\theta = 0$$

because $\frac{\partial\tau}{\partial r}$ is bounded in U, and $\lim_{\epsilon\to 0} \epsilon \log \epsilon = 0$. Q.E.D.

Now suppose $[r_1, r_2]$ is an interval such that all $r \in (r_1, r_2)$ obey ($\mathcal{O}\!\!\mathcal{l}$) and ($\mathcal{L}$) while r_1 and r_2 may violate ($\mathcal{O}\!\!\mathcal{l}$) or ($\mathcal{L}$). We define:

$$\frac{1}{2\pi} \int_{\partial V[t]} f^* u_B {*d\tau} \Big|_{r_1}^{r_2} = \lim_{d\uparrow r_2,\, c\downarrow r_1} \frac{1}{2\pi} \int_{\partial V[t]} f^* u_B {*d\tau} \Big|_c^d$$

With this definition, (4.8) is obviously extended to such

$[r_1, r_2]$ where the end-points r_1, r_2 may violate $(\mathfrak{A})$ or $(\mathfrak{L})$. If $[r_2, r_3]$ is another interval such that each interior r satisfies $(\mathfrak{A})$ and $(\mathfrak{L})$ while r_2, r_3 may not, then Lemma 4.4 implies that

$$\frac{1}{2\pi}\int_{\partial V[t]} f^* u_B * d\tau \Big|_{r_1}^{r_2} + \frac{1}{2\pi}\int_{\partial V[t]} f^* u_B * d\tau \Big|_{r_2}^{r_3}$$
$$= \frac{1}{2\pi}\int_{\partial V[t]} f^* u_B * d\tau \Big|_{r_1}^{r_3}.$$

Now let $[r_0, r_n]$ be an arbitrary sub-interval of $(r(\tau), s)$. The points in $[r_0, r_n]$ at which $(\mathfrak{A})$ or $(\mathfrak{L})$ fails is finite in number, say, $\{r_0, r_1, \ldots, r_n\}$. By the preceding discussion, we have

$$\int_{r_i}^{r_{i+1}} n(t,B)dt + \frac{1}{2\pi}\int_{\partial V[t]} f^* u_B * d\tau \Big|_{r_i}^{r_{i+1}} = \int_{r_i}^{r_{i+1}} v(t)dt$$

for $i = 0, \ldots, n-1$. Add these n equations and the second terms on the left telescope into $\frac{1}{2\pi}\int_{\partial V[t]} f^* u_B * d\tau \Big|_{r_0}^{r_n}$. Thus we have:

$$\int_{r_0}^{r_n} n(t,B)dt + \frac{1}{2\pi}\int_{\partial V[t]} f^* u_B * d\tau \Big|_{r_0}^{r_n} = \int_{r_0}^{r_n} v(t)dt.$$

We are therefore led to the definition of the <u>order function</u>:

$$T(r) = \int_{r_0}^{r} v(t)dt = \frac{1}{\pi}\int_{r_0}^{r} dt \int_{V[t]} f^*\omega$$

and the <u>counting function</u>:

$$N(r,B) = \int_{r_0}^{r} n(t,B)dt.$$

The number r_0 is always assumed to be above $r(\tau)$, but once chosen, it will be fixed. We have thus proved

<u>Theorem 4.5 (FMT)</u>. Let $f\colon V \to G(n,k) \subseteq P_{\ell(k)-1}\mathbb{C}$ be holomorphic, such that $k = 0,\ldots,n-1$, and $f(V)$ does not lie in the polar divisor Σ_B. Suppose V admits a harmonic exhaustion, then for any $r \geq r(\tau)$:

$$N(r,B) + \frac{1}{2\pi}\int_{\partial V[t]} f^* u_B {*d\tau}\Big|_{r_0}^{r} = T(r).$$

As in the case of meromorphic functions, we call the second term on the left the <u>compensating term</u>. What does it compensate for? Let $B^{\perp}$ be the polar space of B, so $B^{\perp}$ is an $(n-k-1)$-dimensional projective subspace of $P_n\mathbb{C}$. By the remarks after (4.4), if each $f(p)$ $(p \in V[r])$ never meets $B^{\perp}$, then $N(r,B) = 0$. The same reasoning shows that if the subset $\bigcup_{p\in V[r]} f(p)$ of $P_n\mathbb{C}$ meets $B^{\perp}$ very rarely, then $N(r,B)$ is small. But $T(r)$ is independent of B, so the above identity implies that the compensating has to be relatively large in this case. Thus the compensating term compensates for the deficiency in the intersection of $B^{\perp}$ with $\bigcup_{p\in V[r]} f(p)$.

Now observe that for any $\Lambda \in \mathbb{C}^{\ell(k)}$, $|\Lambda,B| \leq |\Lambda||B| = |\Lambda|$ ($|B| = 1$ because $B \in G(n,k) \subseteq P_{\ell(k)-1}\mathbb{C}$, and the latter is by definition the quotient space of the unit sphere). So $\log \frac{|\Lambda|}{|\Lambda,B|} \geq 0$. By Theorem 4.1(iii), $u_B \geq 0$. Consequently, because ${*d\tau}$ induces a positive measure on $\partial V[r]$ (since it is coherent with the orientation of $\partial V[r]$), we have

$$(4.10) \qquad \int_{\partial V[r]} f^* u_B * d\tau > 0 \quad \text{for all} \quad r \geq r(\tau).$$

There is one more fact we need before we can derive the basic inequality. This fact is

<u>Lemma 4.6.</u> $\int_{\partial V[r]} f^* u_B * d\tau$ for a fixed r is a continuous function of B.

Let us assume this for a moment and prove the sought for inequality:

(4.11) $N(r,B) < T(r) + \text{const.}$, where the constant is independent of r and B.

For,

$$N(r,B) = T(r) + \frac{1}{2\pi} \int_{\partial V[r_0]} f^* u_B * d\tau - \frac{1}{2\pi} \int_{\partial V[r]} f^* u_B * d\tau$$

$$< T(r) + \frac{1}{2\pi} \int_{\partial V[r_0]} f^* u_B * d\tau. \qquad (4.10)$$

So we may choose the constant to be the maximum of the continuous function $B \mapsto \frac{1}{2\pi} \int_{\partial V[r_0]} f^* u_B * d\tau$ defined on the compact manifold $G(n,k)$. (Lemma 4.6).

*<u>Proof of Lemma 4.6.</u> By virtue of facts (α) and (β) of the proof of Lemma 4.4, it is equivalent to proving the continuity in B of the integral $\int_{V[r]} d(f^* u_B * d\tau)$.

Let $f^{-1}(\Sigma_B) \cap V[r] = \{p_1, \ldots, p_\ell\}$ and let each p_j be surrounded by a coordinate neighborhood on which is defined a fixed coordinate function z_j such that $z_j(p_j) = 0$. Let

$W_j = \{|z_j| \leq \epsilon\}$ and define $W = \bigcup_{j=1}^{\ell} W_j$. Then,

$$\int_{V[r]} d(f^* u_B * d\tau) = \int_{V[r]-W} d(f^* u_B * d\tau) + \Sigma_{j=1}^{\ell} \int_{W_j} d(f^* u_B * d\tau).$$

Now $f(V[r]-W)$ is disjoint from Σ_B, so $f^* u_B$ is a C^∞ function of B on $V[r]-W$ (see (iii) of Theorem 4.1), so there is no question of the continuous dependence of the first integral on B. We only have to examine each summand of the last sum carefully. Fix a j and let y be a reduced representation of f in W_j. By now, it is familiar that $f^* u_B = \log \frac{|y|}{|y,B|}$. So

$$\int_{W_j} d(f^* u_B * d\tau) = \int_{W_j} d(\log |y| * d\tau) - \int_{W_j} d(\log |y,B| * d\tau).$$

Since $|y| > 0$, $\log |y|$ is C^∞ and independent of B. So the first integral of the right side may be left out of consideration. Therefore what we must prove is the following: let B_j be a sequence of projective k spaces in $P_n\mathbb{C}$ converging to B (in the sense that we can pick representatives of B_j and B in $\mathbb{C}^{\ell(k)}$ so that the coefficients of the representatives of B_j converge individually to those of B), then

$$\int_{W_j} d(\log |y,B_j| * d\tau) \to \int_{W_j} d(\log |y,B| * d\tau).$$

Now recall that $y(a_j) \in \Sigma_B$, so the holomorphic function $\langle y,B\rangle$ has a zero at a_j. For convenience, we shall also assume W_j is so small that a_j is the only zero of $\langle y,B\rangle$ in W_j. Furthermore, it is obvious that $\langle y,B_j\rangle$ converges uniformly to $\langle y,B\rangle$ on W_j. To prove the above (and hence the lemma), it therefore suffices to prove the following:

Let $\{g_j\}$ be a sequence of holomorphic functions defined on the closed unit disc Δ and converge uniformly on Δ to g, and let φ be a C^∞ one-form on Δ. Assume that $g(0) = 0$ and vanishes nowhere else. Then if Δ' is the closed disc of radius $\frac{1}{2}$ about the origin,

$$\int_{\Delta'} d(\log |g_j|\varphi) \to \int_{\Delta'} d(\log |g|\varphi)$$

Now the left side equals $\int_{\Delta'} \frac{1}{|g_j|} d|g_j| \wedge \varphi + \int_{\Delta} \log |g_j| d\varphi$. So it is equivalent to proving:

$$\int_{\Delta'} \frac{1}{|g_j|} d|g_j| \wedge \varphi \to \int_{\Delta'} \frac{1}{|g|} d|g| \wedge \varphi. \tag{4.12}$$

$$\int_{\Delta'} \log |g_j| d\varphi \to \int_{\Delta'} \log |g| d\varphi. \tag{4.13}$$

By assumption, there is a positive integer m so that $g(z) = z^m h(z)$, where h is holomorphic in Δ and has no zero in Δ. By Hurwitz's theorem, for each j, there are points $a_{j_1},\ldots,a_{j_m}$ of Δ (possibly not all distinct) such that $a_{j_1} \to 0,\ldots,a_{j_m} \to 0$ and $g_j(z) = (z-a_{j_1})\cdots(z-a_{j_m})h_j(z)$, where h_j is free of zeroes in Δ. To make the notation a little simpler, let us assume that $a_{j_1} = \cdots = a_{j_m}$, and we simply call it a_j. The reader will perceive that this simplification by no means restricts the generality of the subsequent discussion. So we have $g_j(z) = (z-a_j)^m h_j(z)$, $a_j \to 0$ and h_j never zero in Δ. We now claim that h_j converges uniformly to h in Δ'. To begin with, we may assume that all a_j are in the interior of Δ'. For every $\zeta \in \Delta'$,

$$h_j(\zeta) = \frac{1}{2\pi\sqrt{-1}} \int_{\partial\Delta} \frac{h_j(z)dz}{z - \zeta} = \frac{1}{2\pi\sqrt{-1}} \int_{\partial\Delta} \frac{g_j(z)}{(z-a_j)^m(z-\zeta)}\, dz,$$

$$h(\zeta) = \frac{1}{2\pi\sqrt{-1}} \int_{\partial\Delta} \frac{h(z)dz}{z-\zeta} = \frac{1}{2\pi\sqrt{-1}} \int_{\partial\Delta} \frac{g(z)}{z^m(z-\zeta)}\, dz.$$

Since the integrand of the integral of $h_j(\zeta)$ converges uniformly to the integrand of the integral of $h(\zeta)$ on $\partial\Delta$, we have proved our claim.

(4.12) now reads: for $a_j \to 0$, $h_j \to h$,

$$\int_{\Delta'} \frac{m}{|z-a_j|}\, d|z-a_j| \wedge \varphi + \int_{\Delta'} \frac{1}{|h_j|}\, d|h_j| \wedge \varphi$$
$$\to \int_{\Delta'} \frac{m}{|z|}\, d|z| \wedge \varphi + \int_{\Delta'} \frac{1}{|h|}\, d|h| \wedge \varphi$$

Since h_j and h are zero-free and furthermore, h_j and all of its derivatives converge uniformly to h on Δ, it is obvious that the second integral on the left converges to the second integral on the right. So to prove (4.12), it suffices to prove: for $a_j \to 0$,

$$\int_{\Delta'} \frac{1}{|z-a_j|}\, d|z-a_j| \wedge \varphi \to \int_{\Delta'} \frac{1}{|z|}\, d|z| \wedge \varphi.$$

Let $a_j = \alpha_j + \sqrt{-1}\beta_j$. The above simplifies to:

$$\int_{\Delta'} \frac{(x-\alpha_j)dx \wedge \varphi + (y-\beta_j)dy \wedge \varphi}{|z-a_j|^2} \to \int_{\Delta'} \frac{xdx \wedge \varphi + ydy \wedge \varphi}{|z|^2}.$$

To prove this, it is clearly sufficient to prove the following: let f be a C^∞ function on Δ', then $\alpha_j + \sqrt{-1}\beta_j \equiv a_j \to 0$ implies

$$(4.14)\quad \begin{cases} \displaystyle\int_{\Delta'} \frac{(x-\alpha_j)}{|z-a_j|^2}\, f dx\wedge dy \to \int_{\Delta'} \frac{x}{|z|^2}\, f dx\wedge dy, \\ \displaystyle\int_{\Delta'} \frac{(y-\beta_j)}{|z-a_j|^2}\, f dx\wedge dy \to \int_{\Delta'} \frac{y}{|z|^2}\, f dx\wedge dy. \end{cases}$$

Let us prove the first one, say. Let Δ_j be the disc of radius $\frac{1}{2}$ about a_j and let χ_j be the characteristic function of $\Delta_j \cap \Delta'$. Define $f_j(\zeta) = f(\zeta + a_j)$, $\tilde{\chi}_j(\zeta) = \chi_j(\zeta+a_j)$, and let $\mathbb{C}$ denote the complex plane as usual. Then,

$$\int_{\Delta'} f \frac{(x-\alpha_j)}{|z-a_j|^2}\, dxdy = \int_{\mathbb{C}} \tilde{\chi}_j f_j \frac{x}{|z|^2}\, dxdy + \epsilon_j$$

where obviously $\epsilon_j \to 0$ as $a_j \to 0$. In view of Lebesgue's bounded convergence theorem,

$$\int_{\mathbb{C}} \tilde{\chi}_j f_j \frac{x}{|z|^2}\, dxdy \to \int_{\Delta'} \frac{x}{|z|}\, fdxdy$$

and (4.14) is proved. It remains to prove (4.13). Using $g = z^m h$ and $g_j = (z-a_j)^m h_j$, (4.13) becomes

$$\int_{\Delta'} \log |h_j|d\varphi + m\int_{\Delta'} \log |z-a_j|d\varphi \to \int_{\Delta'} \log |h|d\varphi + m\int_{\Delta'} \log |z|d\varphi.$$

Since h_j converges uniformly to h on Δ and both are zero-free in Δ, the first integral on the left clearly converges to the first integral on the right. So it suffices to prove: $a_j \to 0$ implies

$$\int_{\Delta'} \log |z-a_j|d\varphi \to \int_{\Delta'} \log |z|d\varphi.$$

But the method of proof of (4.14) applies equally well to this situation, so the lemma is completely proved. Q.E.D.

§3. In this section, we specialize Theorem 4.5 to the associated curve of rank k (Chapter III, §3) to obtain various refinements. Recall that we assumed our original holomorphic curve $x: V \to P_n\mathbb{C}$ to be nondegenerate. By Lemma 3.6, ${}_kx(V)$ does not lie in any polar divisor Σ_B of $G(n,k)$. Theorem 4.5 therefore implies that for every $A^k \in G(n,k)$,

$$N_k(r,A^k) + \frac{1}{2\pi}\int_{\partial V[t]} {}_kx^* u_A{}^* d\tau \Big|_{r_0}^{r} = T_k(r)$$

where we have attached a subscript k to both the counting function N and the order function T to distinguish their rank. We propose to simplify the compensating term. Recall that r_0 and r are both above $r(\tau)$, so the line integral of the compensating term is taking place in the domain of harmonicity of τ. But there, we can use the special coordinate function $\sigma = \tau + \sqrt{-1}\rho$ (Lemma 2.4 and the remarks after Definition 2.2) except at the critical points of τ. We can therefore define X_σ^k as in (3.2) outside of the critical points of τ. As noted before, σ is defined only up to a translation in the imaginary part ρ, but (3.3) shows that X_σ^k is well-defined despite this ambiguity. The mapping $p \mapsto X_\sigma^k(p)$ into $C^{\ell(k)} \cong \Lambda^{k+1}\mathbb{C}^{n+1}$ we still denote by X_σ^k and by (3.4), ${}_k\pi \circ X_\sigma^k = {}_kx$ makes sense except on a discrete set of points in $V - V[r(\tau)]$, (this discrete set being the union of the zeroes of X_σ^k and the critical points of τ). Since integration always ignores finite point sets, the following is therefore valid:

$$\int_{\partial V[t]} {}_k x^* u_A * d\tau = \int_{\partial V[t]} ({}_k\pi \circ X_\sigma^k)^* u_A * d\tau$$
$$= \int_{\partial V[t]} \log \frac{|X_\sigma^k|}{|X_\sigma^k, A^k|} * d\tau$$

by virtue of Theorem 4.1 (iii). So if we define:

$$(4.15) \qquad m_k(r, A^k) = \frac{1}{2\pi} \int_{\partial V[t]} \log \frac{|X_\sigma^k|}{|X_\sigma^k, A^k|} * d\tau \Big|_{r_0}^{r}$$

then we obtain:

$$(4.16) \qquad N_k(r, A^k) + m_k(r, A^k) = T_k(r),$$

which holds for all $A^k \in G(n,k)$ if x is nondegenerate.

Our next task is to derive two other expressions for $T_k(r)$. Let U be a coordinate neighborhood in V on which is a coordinate function z. Let X_z^k be defined as in (3.2), then (4.2) and (3.4) imply immediately that outside the zeroes of X_z^k:

$$(4.17) \qquad {}_k x^* \omega = \frac{1}{2} dd^c \log |X_z^k| .$$

Since $T_k(r) = \frac{1}{\pi} \int_{r_0}^{r} dt \int_{V[t]} {}_k x^* \omega$, we may write in rather suggestive notation (but not-too-correctly) that $T_k(r) = \frac{1}{2\pi} \int_{r_0}^{r} dt \int_{V[t]} dd^c \log |X_z^k|$.

Still keeping the same notation as above, we obtain from (4.1) and ${}_k\pi \circ X_z^k = {}_k x$ the following: outside the zeroes of X_z^k,

$${}_k x^* \omega = \frac{\sqrt{-1}}{2} \frac{1}{\langle X_z^k, X_z^k \rangle^2} \{ \langle X_z^k, X_z^k \rangle \langle dX_z^k, dX_z^k \rangle - \langle dX_z^k, X_z^k \rangle \langle X_z^k, dX_z^k \rangle \}$$

Now by (3.2), $X_z^k = \mathfrak{X} \wedge \mathfrak{X}^{(1)} \wedge \cdots \wedge \mathfrak{X}^{(k)}$ and so

$$dX_z^k = \tilde{x}\wedge\tilde{x}^{(1)}\wedge\ldots\wedge\tilde{x}^{(k-1)}\wedge\tilde{x}^{(k+1)}dz = \{X_z^{(k-1)}\wedge\tilde{x}^{(k+1)}\}dz$$

for $k = 1,\ldots,n-1$. To extend the validity to $k = 0$, we simply define:

$$X_z^{-1} \equiv 1.$$

Thus on U minus the zeroes of X_z^k:

$$\begin{aligned} {}_kx^*\omega = \frac{\sqrt{-1}}{2}\,\frac{1}{\langle X_z^k,X_z^k\rangle^2}\{&\langle X_z^k,X_z^k\rangle\langle X_z^{(k-1)}\wedge\tilde{x}^{(k+1)},X_z^{(k-1)}\wedge\tilde{x}^{(k+1)}\rangle \\ &- \langle X_z^{(k-1)}\wedge\tilde{x}^{(k+1)},X_z^k\rangle\langle X_z^k,X_z^{(k-1)}\wedge\tilde{x}^{(k+1)}\rangle\}dz\wedge d\bar{z}. \end{aligned}$$

Taking into account of the fact that $X_z^k = X_z^{(k-1)}\wedge\tilde{x}^{(k)}$, we may apply Sylvester's theorem on compound determinants (3.8) to conclude: outside the zeroes of X_z^k

$$(4.18)\qquad {}_kx^*\omega = \frac{\sqrt{-1}}{2}\,\frac{|X_z^{k-1}|^2|X_z^{k+1}|^2}{|X_z^k|^4}\,dz\wedge d\bar{z}$$

where we have written as usual, $|X_z^{k-1}|^2$, etc. for $\langle X_z^{k-1},X_z^{k-1}\rangle$, etc. Again, it is tempting to write that

$$T_k(r) = \frac{\sqrt{-1}}{2\pi}\int_{r_0}^{r} dt \int_{V[t]} \frac{|X_z^{k-1}|^2|X_z^{k+1}|^2}{|X_z^k|^4}\,dz\wedge d\bar{z}.$$

The trouble with this, as is the trouble with $T_k(r)$ $= \frac{1}{2\pi}\int_{r_0}^{r} dt \int_{V[t]} dd^c \log|X_z^k|$, is that z is not a globally defined coordinate function, so that the integrand does not make sense on all of V.

This suggests that we should look for a function on V

which can serve as a coordinate function at every point of V. Such a function is provided for by a theorem of Gunning and Narasimhan:

Theorem 4.7 [3]. On every open Riemann surface, there is a holomorphic function whose differential vanishes nowhere.

Let us seize such a function γ on V and fix the notation once and for all. Thus in every sufficiently small open subset U of V, the restriction of γ to U is a coordinate function. Then X_γ^k makes global sense on V and (3.4) implies that the following diagram is commutative:

(4.19)
$$\begin{array}{ccc} & \overset{X_\gamma^k}{\nearrow} & \mathbb{C}^{\ell(k)} - \{0\} \\ & & \downarrow {}_k\pi \\ V' & \xrightarrow{{}_kx} & G(n,k) \subseteq P_{\ell(k)-1}\mathbb{C} \end{array}$$

where V' is the complement of the zeroes of X_γ^k in V. Furthermore, by virtue of (4.17) and (4.18), we now have

(4.20)
$$T_k(r) = \frac{1}{2\pi}\int_{r_o}^{r} dt \int_{V[t]} dd^c \log |X_\gamma^k|$$
$$= \frac{\sqrt{-1}}{2\pi}\int_{r_o}^{r} dt \int_{V[t]} \frac{|X_\gamma^{k-1}|^2 |X_\gamma^{k+1}|^2}{|X_\gamma^k|^2}\, d\gamma \wedge d\bar{\gamma}$$

Consider the first expression of $T_k(r)$ in (4.20). We will apply Stokes' theorem to the integrand $\int_{V[t]} dd^c \log |X_\gamma^k|$ in exactly the same way as we did in Theorem 4.2. In greater detail, X_γ^k will vanish in a finite number of points $\{p_1,\ldots,p_m\} \subseteq V[t]$. So we enclose each p_j by a small

ϵ-ball W_j. Let $W = \bigcup_{j=1}^{m} W_j$. On $V[t]-W$, we can apply Stokes' theorem to the C^∞ form $dd^c \log |X_\gamma^k|$. Then take the limit of the integral as $\epsilon \to 0$. The result is: if the zeroes of $|X_\gamma^k|$ do not fall on $\partial V[t]$, then:

$$(4.21) \quad \frac{1}{2\pi} \int_{V[t]} dd^c \log |X_\gamma^k| = \frac{1}{2\pi} \int_{\partial V[t]} d^c \log |X_\gamma^k| - n_k(t)$$

where by definition:

(4.22) $n_k(t)$ = the sum of the orders of zeroes of $|X_\gamma^k|$ in $V[t]$.

Note that (4.22) makes sense because each component of X_γ^k is a holomorphic function on V. Now repeating the proof of Lemma 2.5 almost word for word, we can show that once $t \geq r(\tau)$ and $\partial V[t]$ contains no zero of $|X_\gamma^k|$, then

$$\frac{1}{2\pi} \int_{\partial V[t]} d^c \log |X_\gamma^k| = \frac{d}{dt}\left(\frac{1}{2\pi} \int_{\partial V[t]} \log |X_\gamma^k|\right).$$

Substitute this into (4.21) and integrate, we get:

$$(4.23) \quad \frac{1}{2\pi} \int_{r_0}^{r} dt \int_{V[t]} dd^c \log |X_\gamma^k| = \frac{1}{2\pi} \int_{\partial V[t]} \log |X_\gamma^k| * d\tau \Big|_{r_0}^{r} - N_k(r)$$

where we have written $N_k(r) = \int_{r_0}^{r} n_k(t)dt$. (4.23) is only true if every $t \in [r_0, r]$ has the property that $\partial V[t]$ contains none of the zeroes of $|X_\gamma^k|$. But now the analogue of Lemma 4.4 is valid; again the proof can be transferred to this case almost verbatim. Therefore $\int_{\partial V[t]} \log |X_\gamma^k| * d\tau$ is a continuous function of t. The standard arguments that led to Theorem 2.7 and Theorem 4.5 now show that (4.23) is valid for any subinterval

$[r_0, r]$ of $(r(\tau), s)$. Consequently, (4.20) implies that

$$(4.24) \qquad T_k(r) = \frac{1}{2\pi} \int_{\partial V[t]} \log |X_\gamma^k| * d\tau \Big|_{r_0}^{r} - N_k(r).$$

We now summarize (4.16)-(4.18) and (4.24). Recall first the various definitions.

$$N_k(r, A^k) = \int_{r_0}^{r} n_k(t, A^k)\,dt$$

where $n_k(t, A^k)$ = sum of the orders of zeroes of $\langle {}_k x, A^k \rangle$ in $V[t]$.

$$N_k(r) = \int_{r_0}^{r} n_k(t)\,dt$$

where $n_k(t)$ = sum of the orders of zeroes of the function $|X_\gamma^k|$ in $V[t]$, where γ is a fixed function enjoying the property of Theorem 4.7.

$$T_k(r) = \frac{1}{\pi} \int_{r_0}^{r} v_k(t)\,dt$$

where $v_k(t) = \int_{V[t]} {}_k x^* \omega$, and

$$m_k(r, A^k) = \frac{1}{2\pi} \int_{\partial V[t]} \log \frac{|X_\sigma^k|}{|X_\sigma^k, A^k|} * d\tau \Big|_{r_0}^{r},$$

where σ is a holomorphic function having τ as its real part.

<u>Theorem 4.8 (FMT of rank k)</u>. Let $x\colon V \to P_n\mathbb{C}$ be a non-degenerate holomorphic curve and let ${}_k x\colon V \to G(n,k) \subseteq P_{\ell(k)-1}\mathbb{C}$ be its associated holomorphic curve of rank k, $k = 0, \ldots, n-1$. Let V admit a harmonic exhaustion. Then for each $A^k \in G(n,k)$ and for $r \geq r(\tau)$:

$$N_k(r,A^k) + m_k(r,A^k) = T_k(r)$$

If γ is the fixed function on V having the property of Theorem 4.7, then

$$T_k(r) = \frac{1}{2\pi}\int_{\partial V[t]} \log |X_\gamma^k| * d\tau \Big|_{r_o}^{r} - N_k(r).$$

Furthermore, let U be a coordinate neighborhood in V with coordinate function z, then except on a discrete point set:

$$\begin{aligned} {}_k x^* \omega &= \frac{1}{2}\, dd^c \log |X_z^k| \\ &= \frac{\sqrt{-1}}{2} \frac{|X_z^{k-1}|^2 |X_z^{k+1}|^2}{|X_z^k|^4}\, dz \wedge d\bar{z} \end{aligned}$$

where $X_z^{-1} \equiv 1$ by definition.

<u>Remarks</u>. (1) In the same setting, we may restate (4.11) as follows:

(4.25) $N_k(r,A^k) < T_k(r) + c_k$, where c_k is a constant independent of r and A^k.

(2) The case $k = 0$ is notable for its simplicity, so we state the above for this case separately. For any $a \in P_n\mathbb{C}$,

$$\begin{aligned} m_0(r,a) &= \frac{1}{2\pi}\int_{\partial V[t]} \log \frac{|\tilde{x}|}{|\tilde{x},a|} * d\tau \Big|_{r_o}^{r} \\ T_0(r) &= \frac{1}{2\pi}\int_{\partial V[t]} \log |\tilde{x}| * d\tau \Big|_{r_o}^{r} - N_0(r) \\ x^*\omega &= \frac{1}{2}\, dd^c \log |\tilde{x}| \\ &= \frac{\sqrt{-1}}{2} \frac{|\tilde{x}\wedge\tilde{x}^{(1)}|^2}{|\tilde{x}|^4}\, dz \wedge d\bar{z}\ . \end{aligned}$$

(3) We have introduced the holomorphic function γ on V and with it, the mapping $X^k_\gamma\colon V \to \mathbb{C}^{\ell(k)}$ which induces ${}_kx\colon V \to P_{\ell(k)-1}\mathbb{C}$. Using X^k_γ, we now give an equivalent definition of $n_k(t,A^k)$:

(4.25a) $n_k(t,A^k)$ = sum of the orders of zeroes of the function $\dfrac{|X^k_\gamma,A^k|}{|X^k_\gamma|}$ in $V[t]$.

Take a $p \in V$; if $X^k_\gamma(p) \neq 0$, then obviously the order of zero at p of the quotient equals the order of zeroes of $|X^k_\gamma,A^k|$, equals the order of zero at p of $\langle X^k_\gamma,A^k\rangle$. But in a small neighborhood of p, ${}_k\pi \circ X^k_\gamma = {}_kx$ by (3.4), so each $X^k_\gamma(q)$ is a representative of ${}_kx(q)$ for all q in this neighborhood, which in turn simply means the order of zero of $\langle X^k_\gamma,A^k\rangle$ at p is the order of zero of $\langle {}_kx,A^k\rangle$ at p. Next, let $X^k_\gamma(p) = 0$, then in a sufficiently small neighborhood U of p, we can write $X^k_\gamma = z^mY$, where z is a p-centered coordinate function on U and Y is such that it is never zero on U. Then $Y\colon U \to \mathbb{C}^{\ell(k)} - \{0\}$ is a reduced representation of ${}_kx$ in U (this is obvious) so that ${}_k\pi \circ Y = {}_kx$. Thus each $Y(q)$ is a representative of each ${}_kx(q)$, $q \in U$, and so the order of zero at p of $\langle {}_kx,A^k\rangle$ is just the order of zero at p of $\langle Y,A^k\rangle$. The latter is equal to the order of zero at p of $|Y,A^k|$, and hence equal to the order of zero of $\frac{1}{|Y|}|Y,A^k|$ at p (because Y is never zero in U). Since $|X^k_\gamma| = |z^m||Y|$, it follows that $\frac{1}{|Y|}|Y,A^k| = \frac{1}{|X^k_\gamma|}|X^k_\gamma,A^k|$ and we have again proved that the order of zero at p of $\frac{1}{|X^k_\gamma|}|X^k_\gamma, A^k|$ is equal to the order of zero at p of $\langle_k x, A^k\rangle$. According to the definition of $n_k(t,A^k)$ given before Theorem 4.8, we have proved (4.25).

(4) Although only incidental to our purpose, we nevertheless give an extension of the integral formula (2.23). As noted in Chapter I §2, the Fubini-Study metric of $P_{\ell(k)-1}\mathbb{C}$ induces a Kahler structure on $G(n,k)$. The volume form of this Kahler metric on $G(n,k)$ we shall denote genericall by Ω. If now g is an integrable function on $G(n,k)$, we define the arithmetic mean of g to be:

$$(4.26)\qquad \mathfrak{M}_{G(n,k)}\, g \equiv \left(\int_{G(n,k)} \Omega\right)^{-1}\left(\int_{G(n,k)} g\Omega\right)$$

Sometimes we write also $\mathfrak{M}_{A^k}\, g(A^k)$ to emphasize the variable of integration. The main observation here is:

$$(4.27)\qquad \mathfrak{M}_{A^k}\, N_k(r,A^k) = T_k(r)$$

According to (4.16), this is equivalent to:

$$(4.28)\qquad \mathfrak{M}_{A^k}\, m_k(r,A^k) = 0.$$

To prove this, first note that if B is an arbitrary member of $G(n,k)$, the function $g(A^k) = \log\frac{|B|}{|B,A^k|}$ is nonnegative and measurable on $G(n,k)$. We have noted the nonnegativeness of g on previous occasions; it is measurable because it is continuous in $G(n,k) - \Sigma_B$. Hence for some μ, $0 < \mu \leq \infty$,

$$\mathfrak{M}_{A^k} \log \frac{|B|}{|B,A^k|} = \mu.$$

Our first claim is that μ is independent of B. This is due to two facts: (i) $U(n+1)$ acting on the function $\log \frac{|B|}{|B,A^k|}$ as a transformation group of $G(n,k)$ leaves the function invariant. (ii) $U(n+1)$ is a transitive group of isometries of $G(n,k)$ (Lemma 1.2) and in particular, leaves Ω invariant. Next we claim that $\mu < \infty$. For Lemma 4.6 says that $\int_{\partial V[t]} \log \frac{|X_\sigma^k|}{|X_\sigma^k, A^k|} *d\tau$ is a continuous function of A^k and therefore

$$0 < \mathfrak{M}_{A^k} \int_{\partial V[t]} \frac{|X_\sigma^k|}{|X_\sigma^k, A^k|} *d\tau < \infty .$$

But $\log \frac{|X_\sigma^k|}{|X_\sigma^k, A^k|}$ is nonnegative and measurable on $\partial V[t] \times G(n,k)$ (again because it is continuous except on a set of measure zero), so Fubini's theorem gives $\mathfrak{M}_{A^k} \int_{\partial V[t]} = \int_{\partial V[t]} \mathfrak{M}_{A^k}$. Using the first claim and the corollary to Lemma 2.4, we deduce that the above inequalities may be rewritten as

$$0 < \mu L < \infty .$$

Since $0 < L < \infty$, we conclude that $\mu < \infty$. Therefore, by (4.15),

$$\mathfrak{M}_{A^k} m_k(r,A^k) = \mathfrak{M}_{A^k} \frac{1}{2\pi} \int_{\partial V[t]} \log \frac{|X_\sigma^k|}{|X_\sigma^k, A^k|} *d\tau \Big|_{r_o}^{r}$$

$$= \frac{1}{2\pi} \int_{\partial V[t]} (\mathfrak{M}_{A^k} \log \frac{|X_\sigma^k|}{|X_\sigma^k, A^k|})^* d\tau \Big|_{r_o}^{r}$$

$$= \frac{1}{2\pi} \int_{\partial V[t]} \mu^* d\tau \Big|_{r_o}^{r}$$

$$= \frac{1}{2\pi}(\mu L - \mu L)$$

$$= 0,$$

thereby proving (4.28).

§4. The FMT of rank k introduces the quantity $m_k(r, A^k)$ as given by (4.15). Since by definition $\langle X_\sigma^k, A^k \rangle = A^k \rfloor X_\sigma^k$, we may rewrite (4.15) as

$$m_k(r, A^k) = \frac{1}{2\pi} \int_{\partial V[t]} \log \frac{|X_\sigma^k|}{|X_\sigma^k \rfloor A^k|} *d\tau \Big|_{r_o}^{r} .$$

Now let h be an integer obeying $-1 \leq h \leq n$ and let A^h be a unit decomposable (h+1)-vector. In Chapter V, we shall encounter expressions of the type:

$$(4.29) \qquad m_k(r, A^h) \equiv \frac{1}{2\pi} \int_{\partial V[t]} \log \frac{|X_\sigma^k|}{|X_\sigma^k \rfloor A^h|} *d\tau \Big|_{r_o}^{r} ,$$

and the purpose of this section is to give an interpretation of this. Invoking the distinguished function γ on V, we may write as a consequence of (3.3) that

$$m_k(r, A^h) = \frac{1}{2\pi} \int_{\partial V[t]} \log \frac{|X_\gamma^k|}{|X_\gamma^k \rfloor A^h|} *d\tau \Big|_{r_o}^{r} .$$

Recalling (4.24), we have,

$$T_k(r) = \frac{1}{2\pi} \int_{\partial V[t]} \log |X_\gamma^k| * d\tau \Big|_{r_0}^{r} - N_k(r).$$

Now introduce

$n_k(t,A^h)$ = sum of the orders of zeroes of the function $\dfrac{|X_\gamma^k \lrcorner A^h|}{|X_\gamma^k|}$ in $V[t]$

and write $N_k(r,A^h) = \int_{r_0}^{r} n_k(t,A^h)dt$. When $h = k$, (4.25) shows that this $N_k(r,A^k)$ coincides with the usual counting function. We further define:

$$T_k(r,A^h) = \frac{1}{2\pi} \int_{\partial V[t]} \log |X_\gamma^k \lrcorner A^h| * d\tau \Big|_{r_0}^{r} - N_k(r,A^h) - N_k(r).$$

Then upon adding the various terms, we find

$$m_k(r,A^h) + T_k(r,A^h) + N_k(r,A^h) = T_k(r) \tag{4.30}$$

Since the FMT says that for $h = k$, $m_k(r,A^k) + N_k(r,A^k) = T_k(r)$, (4.30) leads to:

$$T_k(r,A^k) = 0. \tag{4.31}$$

We now wish to give a geometric interpretation of the formal object $T_k(r,A^h)$. Let us define

$\tilde{n}_k(t,A^h)$ = sum of the orders of zeroes of the function $|X_\gamma^k \lrcorner A^h|$ in $V[t]$,

then as a matter of comparing definitions, we see that $n_k(t,A^h) + n_k(t) = \tilde{n}_k(t,A^h)$. Hence $N_k(t,A^h) + N_k(t) = \tilde{N}_k(t,A^h)$, where

$\tilde{N}_k(t,A^h) = \int_{r_0}^{r} \tilde{n}_k(t,A^h)dt$. So we may rewrite $T_k(r,A^h)$ as:

$$(4.32) \qquad T_k(r,A^h) = \frac{1}{2\pi} \int_{\partial V[t]} \log |X_\gamma^k \rfloor A^h| * d\tau \Big|_{r_0}^{r} - \tilde{N}_k(r,A^h).$$

Keeping (4.32) in mind, we separate our considerations into two cases:

Case 1: $k < h$. In Chapter III §4, we introduced the projection curve of x into A^h, which was denoted by ${}_Ax: V \to A^h$. Let E be the $(h+1)$-dimensional vector subspace of $\mathbb{C}^{n+1}$ correspondi[ng] to A, then the mapping ${}_AX_\gamma^k: V \to \Lambda^{k+1}E$ defined in Lemma 3.7 induces the associated curve of rank k of ${}_Ax$. By that lemma, $|{}_AX_\gamma^k| = |X_\gamma^k \rfloor A^h|$, consequently, the interpretation of the order function as given in (4.24) shows that $T_k(r,A^h)$ in this case is exactly the order function of the associated curve of rank k of ${}_Ax$.

We can also look at this from a second point of view. In §6 of Chapter III, we introduced the contracted curve of the second kind, $X^k \rfloor A^h: V \to G(n,h-k-1) \subseteq P_{\ell(h-k-1)-1}\mathbb{C}$. Since $X_\gamma^k \rfloor A^h: V \to \Lambda^{h-k}\mathbb{C}^{n+1}$ clearly induces $X^k \rfloor A^h$, the interpretation in (4.24) of the order function (for $k = 0$) again shows that $T_k(r,A^h)$ is the order function of the holomorphic curve $X^k \rfloor A^h$.

Case 2: $h < k$. We introduced in §5 of Chapter III the contracted curve of the first kind, $X^k \rfloor A^h: V \to G(n,k-h-1) \subseteq P_{\ell(k-h-1)-1}\mathbb{C}$. Obviously, $X_\gamma^k \rfloor A^h: V \to \Lambda^{k-h}\mathbb{C}^{n+1}$ induces $X^k \rfloor A^h$. Since $|A^h \rfloor X_\gamma^k| = |X_\gamma^k \rfloor A^h|$, the interpretation in (4.24) of the order function for $k = 0$ shows that $T_k(r,A^h)$ is the order function of the holomorphic curve $X^k \lrcorner A^h$.

In any case, we have shown that if $h \neq k$, then $T_k(r,A^h)$ is actually the order function of some holomorphic curve. Recall that if $f: V \to P_n\mathbb{C}$ is a holomorphic curve, the order function of f was originally defined to be $\int_{r_0}^r dt \int_{V[t]} f^*\omega$, so the order function is always positive and strictly increasing. Hence

$$(4.33) \qquad T_k(r,A^h) > 0 \quad \text{if} \quad k \neq h.$$

We now summarize the foregoing in the next theorem.

Theorem 4.9. Assumptions as in Theorem 4.8, let A^h be a unit decomposable $(h+1)$-vector, $h = -1,0,\ldots,n$, and define for $k = 0,\ldots,n-1$,

$$m_k(r,A^h) = \frac{1}{2\pi} \int_{\partial V[t]} \log \frac{|X_\sigma^k|}{|X_\sigma^k \lrcorner A^h|} *d\tau \Big|_{r_0}^r$$

$$T_k(r,A^h) = \frac{1}{2\pi} \int_{\partial V[t]} \log |X_\gamma^k \lrcorner A^h| *d\tau - \tilde{N}_k(r,A^h)$$

where $\tilde{N}_k(r,A^h) = \int_{r_0}^r \tilde{n}_k(t,A^h)dt$ and by definition,

$$\tilde{n}_k(t,A^h) = \text{sum of the orders of zeroes of } |X_\gamma^k \lrcorner A^h| \text{ in } V[t].$$

Similarly, let $N_k(r,A^h)$ be defined with the zeroes of $\dfrac{|X_\gamma^k \lrcorner A^h|}{|X_\gamma^k|}$.

Then the following identity holds:

$$m_k(r,A^h) + T_k(r,A^h) + N_k(r,A^h) = T_k(r).$$

Furthermore, $T_k(r,A^k) = 0$, and if $h > k$,

(4.34) $T_k(r,A^h)$ = the order function of the associated holomorphic curve of rank k of the projection curve of x into A^h.
= the order function of the contracted curve of the second kind $X^k \lrcorner A^h$.

and if $h < k$,

(4.35) $T_k(r,A^h)$ = the order function of the contracted curve of the first kind $X^k \lrcorner A^h$.

In connection with the above theorem, there are two basic inequalities which we shall need. We now prove them here.

Lemma 4.10. There exist constants a_k, $k = 0,\dots,n-1$, such that

$$\frac{1}{2\pi}\int_{\partial V[r]} \log \frac{|X_\sigma^k|}{|X_\sigma^k \lrcorner A^h|} *d\tau \le T_k(r) + a_k$$

holds for all r and A^h.

Lemma 4.11. There exist constants b_k, $k = 0,\dots,n-1$, such that

$$T_k(r,A^h) \le T_k(r) + b_k$$

holds for all A^h and r.

Proof of Lemma 4.10. By (4.31) and (4.33) $T_k(r,A^h) \ge 0$. Also, by its very definition, $N_k(r,A^h) > 0$. We therefore deduce from (4.30) that

$$m_k(r,A^h) < T_k(r)$$

which by (4.29) is equivalent to

$$\frac{1}{2\pi}\int_{\partial V[r]} \log \frac{|X_\sigma^k|}{|X_\sigma^k \lrcorner A^h|} *d\tau < T_k(r) + \frac{1}{2\pi}\int_{\partial V[r_0]} \log \frac{|X_\sigma^k|}{|X_\sigma^k \lrcorner A^h|} *d\tau$$

The right-side integral can be proved to be a continuous function of A^h in exactly the same manner that Lemma 4.6 was proved. So we may let a_k be the maximum of this function as A^h varies over $G(n,h)$. Q.E.D.

Proof of Lemma 4.11. We begin by recalling (4.24) and (4.32):

$$T_k(r) = \frac{1}{2\pi}\int_{\partial V[t]} \log |X_\gamma^k| * d\tau \Big|_{r_0}^{r} - N_k(r)$$

$$T_k(r,A^h) = \frac{1}{2\pi}\int_{\partial V[t]} \log |X_\gamma^k \lrcorner A^h| * d\tau \Big|_{r_0}^{r} - \tilde{N}_k(r,A^h)$$

By (1.12), $|X_\gamma^k \lrcorner A^h| \leq |X_\gamma^k||A^h| = |X_\gamma^k|$ (one must remember our notation convention that A^h is a unit decomposable vector!) Hence $\int_{\partial V[r]} \log |X_\gamma^k \lrcorner A^h| * d\tau \leq \int_{\partial V[r]} \log |X_\gamma^k| * d\tau$. Furthermore, every zero of $|X_\gamma^k|$ is of course a zero of $|X_\gamma^k \lrcorner A^h|$, so $N_k(r) \leq \tilde{N}_k(r,A^h)$. Combining these two facts, we have:

$$T_k(r,A^h) + \frac{1}{2\pi}\int_{\partial V[r_0]} \log |X_\gamma^k \lrcorner A^h| * d\tau$$

$$\leq T_k(r) + \frac{1}{2\pi}\int_{\partial V[r_0]} \log |X_\gamma^k| * d\tau$$

or equivalently,

$$T_k(r,A^h) \leq T_k(r) + \frac{1}{2\pi}\int_{\partial V[r_o]} \log \frac{|X_\gamma^k|}{|X_\gamma^k \lrcorner A^h|} * d\tau$$

The situation now is identical with that of the preceding lemma: the last integral is a continuous function of A^h, so we take b_k to be the maximum of this function as A^h wanders over $G(n,h)$. Q.E.D.

§5. We now head towards the Second Main Theorem. It relates the order functions T_{k-1}, T_k, T_{k+1} to the Euler characteristic of $V[r]$. In the first part of the derivation of this formula, however, we need not assume that V is open and that the receiving space is $G(n,k) \subseteq P_{\ell(k)-1}\mathbb{C}$. In fact, the exposition is smoother without these assumptions. So we let M be an n-dimensional hermitian manifold with hermitian metric F whose associated two form we denote by ω. (This redundant usage of F and ω, which have been employed to denote the F-S metric and its Kahler form on $P_n\mathbb{C}$, is intentional). Let $f\colon V \to M$ be a nonconstant holomorphic mapping (V arbitrary) and let D be a compact subdomain on V with C^∞ boundary. We consider the restriction $f\colon D \to M$. df will in general vanish at a finite number of points $\{\alpha_1,\ldots,\alpha_m\}$ of D. We call these the <u>critical points</u> of f in D. If $f(\alpha_j) = p$, let z be an α_j-centered coordinate function and let U be a coordinate neighborhood of p with coordinate map $\zeta\colon U \to O \subseteq \mathbb{C}^n$ such that $\zeta(p) = (0,\ldots,0)$. Then $f^*\zeta$ is a $\mathbb{C}^n$-valued function such that $f^*\zeta(0) = (0,\ldots,0)$ and there will be a positive integer s such that $(z^{-s}df^*\zeta)(0)$

$\neq (0,\dots,0)$ while $(z^{-(s-1)}df^*\zeta)(0) = (0,\dots,0)$. This s is clearly independent of the choice of z and ζ and is therefore an intrinsic invariant of f. We call it the stationary index of f at a_j. We define:

(4.36) $\quad s(D) =$ sum of the stationary indices of f in D.

In $D - \{\alpha_1,\dots,\alpha_m\}$, $f^*F \equiv G$ becomes a hermitian metric. Let K be its Gaussian curvature and Ω its volume form. Of course Ω is also the Kahler form of G and so

(4.37) $$f^*\omega = \Omega \quad \text{in} \quad D - \{\alpha_1,\dots,\alpha_m\}.$$

The nonintegrated form of the Second Main Theorem may now be stated as follows. (The reader may consult at this point the discussion of the Gauss-Bonnet theorem in Chapter II, §3).

Theorem 4.12. Let $f: D \to M$ be a nonconstant holomorphic mapping, where D is a compact subdomain with C^∞ boundary in an arbitrary Riemann surface V and M is an n-dimensional complex manifold with hermitian metric F. Suppose df is nowhere zero on ∂D so that the geodesic curvature κ of ∂D with respect to $G = f^*F$ is defined. Notation as in (4.36) and (4.37), the following holds:

$$2\pi\chi(D) + 2\pi s(D) - \int_{\partial D}\kappa = \int_D K\Omega$$

where $\chi(D)$ denotes the Euler characteristic of D.

Proof. Let $\{\alpha_1,\dots,\alpha_m\}$ be the points where df vanishes and let z_j be a local coordinate function near α_j with

$z_j(\alpha_j) = 0$. Denote by W_j the set $\{|z_j| \leq \epsilon\}$ and let $W = \bigcup_{j=1}^m W_j$. On $D - W$, G is a hermitian metric and we can apply the Gauss-Bonnet theorem:

$$2\pi\chi(D-W) - \int_{\partial D}\kappa + \int_{\partial W}\kappa = \int_{D-W} K\Omega$$

where, for convenience, we also denote by κ the geodesic curvature form of ∂W and the plus sign in front of $\int_{\partial W}\kappa$ is due to the fact that we orient each ∂W_j with respect to W_j (and not with respect to $D-W$). Clearly $\chi(D-W) = \chi(D) - m$ and $\lim_{\epsilon\to 0}\int_{D-W} K\Omega = \int_D K\Omega$. Hence,

$$\int_D K\Omega = 2\pi\chi(D) - \int_{\partial D}\kappa + \lim_{\epsilon\to 0}\Sigma_{j=1}^m\Big(\int_{\partial W_j}\kappa - 2\pi\Big).$$

To prove the theorem, it suffices to show that for each j,

$$(4.38)\qquad \lim_{\epsilon\to 0}\Big(\frac{1}{2\pi}\int_{\partial W_j}\kappa - 1\Big) = \text{stationary index of } f \text{ at } \alpha_j.$$

For this, we explicitly compute κ of ∂W_j. For simplicity, we denote z_j by z. So z is a coordinate function centered at α_j. Let $z = re^{\sqrt{-1}\theta}$ and let $\eta = \log r$, then $\xi = \eta + \sqrt{-1}\theta$ is a multi-valued holomorphic function in $W_j - \{0\}$ and $\partial W_j = \{p\colon \eta(p) = \log\epsilon\}$. In W_j minus a radial slit, ξ is a coordinate function and, we can write (see (2.11)):

$$G = g(d\eta^2 + d\theta^2),$$

$$\Omega = g\, d\eta\wedge d\theta.$$

By (2.14), the geodesic curvature form of ∂W_j has the expression:

(4.39) $$\kappa = \frac{1}{2}\frac{\partial \log g}{\partial \Pi}\, d\theta$$

It remains to find out what g is. Let $f(\alpha_j) = p$ and let $\zeta_1,\ldots,\zeta_n$ be local coordinates at p so that $\zeta_j(p) = 0$, $j = 1,\ldots,n$. Around p, we may write

$$\omega = \frac{\sqrt{-1}}{2}\Sigma_{ij}\, a_{ij}\, d\zeta_i \wedge d\bar{\zeta}_j$$

where ω is the associated two form of F and $\{a_{ij}\}$ is hermitian positive definite. Relative to z and $(\zeta_1,\ldots,\zeta_n)$, f decomposes into n holomorphic functions $f = (f_1,\ldots,f_n)$ so that

$$f^*\omega = \left(\frac{\sqrt{-1}}{2}\Sigma_{ij}\, f^*a_{ij}\frac{\partial f_i}{\partial z}\overline{\frac{\partial f_j}{\partial z}}\right)dz\wedge d\bar{z}\ .$$

Because $f(\alpha_j) = p$, all f_j vanish at 0. Let f_1 have the smallest order of zero among $f_1,\ldots,f_n$, say, and let

$$f_1(z) = z^{\delta}h_1(z)$$

such that h is holomorphic and $h(0) \neq 0$. It is clear that the stationary index of f at α_j is $(\delta-1)$. Similarly let $f_i(z) = z^{\delta}h_i(z)$, where h_i is holomorphic in W_j, $i = 2,\ldots,n$. Then

$$\frac{\partial f_k}{\partial z} = z^{\delta-1}h_k + z^{\delta}\frac{\partial h_k}{\partial z} = z^{\delta-1}\left(h_k + z\frac{\partial h_k}{\partial z}\right)$$

for $k = 1,\ldots,n$. Let $b_k = h_k + z\dfrac{\partial h_k}{\partial z}$. Since $b_1(0) = h_1(0) \neq 0$, b_1 is never zero in a neighborhood of 0. We may clearly assume that the holomorphic function b_1 is never zero in W_j. Let b denote the row vector of functions: $b = [b_1\cdots b_n]$, and let A denote the hermitian positive definite matrix $\{f^*a_{ij}\}$. Then

$$f^*\omega = |z|^{2(\delta-1)}(bA\bar{b}^t)dz \wedge d\bar{z} .$$

Clearly $bA\bar{b}^t$ is a C^∞ function which never vanishes in W_j because b is never zero there and A is positive definite. Furthermore a simple computation shows that in $W_j - \{0\}$, $dz \wedge d\bar{z} = |z|^2 d\eta \wedge d\theta$ (recall that $z = re^{\sqrt{-1}\theta}$ and $\eta = \log r$), so

$$f^*\omega = |z|^{2\delta}(bA\bar{b}^t)d\eta \wedge d\theta.$$

Since $\Omega = f^*\omega$ by (4.37) and $\Omega = g\, d\eta \wedge d\theta$ by definition of g, we see that $g = |z|^{2\delta}(bA\bar{b}^t)$. Consequently, by (4.39),

$$\begin{aligned}\kappa &= \frac{1}{2}\,\frac{\partial \log(|z|^{2\delta}(bA\bar{b}^t))}{\partial\eta}\, d\theta \\ &= \delta\, \frac{\partial \log |z|}{\partial\eta}\, d\theta + \frac{1}{2}\,\frac{\partial \log bA\bar{b}^t}{\partial\eta}\, d\theta \\ &= \delta\, d\theta + C^\infty \text{ form}\end{aligned}$$

because $\eta = \log |z|$ and because $bA\bar{b}^t$ is C^∞ and zero-free. Hence:

$$\lim_{\epsilon\to 0}\Big(\frac{1}{2\pi}\int_{\partial W_j}\kappa - 1\Big) = \delta - 1 .$$

As remarked above, the stationary index of f at α_j is $(\delta-1)$, so this proves (4.38). Q.E.D.

The rest of this section constitutes a digression and may be omitted without loss of continuity. We would like to elaborate on Theorems 4.2 and 4.12 for the case of a compact Riemann surface D <u>without</u> boundary. So let us fix a holomorphic mapping $x: M \to P_n\mathbb{C}$ with the explicit assumptions that (i) x is nondegenerate (i.e., the only subspace of $P_n\mathbb{C}$ containing $x(M)$ is $P_n\mathbb{C}$ itself) and (ii) M is a compact Riemann surface without boundary. We would like to define the associated curve of rank k of x, i.e., ${}_kx: M \to G(n,k) \subseteq P_{\ell(k)-1}\mathbb{C}$. The definition given in §3 of Chapter III does not apply because it made use of a map $\tilde{x}: M \to \mathbb{C}^{n+1}$ which induces x. Since x is compact, no such $\tilde{x}$ exists, so we must modify the definition somewhat. In the formula (3.2) which defines X_z^k in a coordinate neighborhood U, we replace $\tilde{x}$ throughout by an arbitrary reduced representation $x^*: U \to \mathbb{C}^{n+1} - \{0\}$ of x in U. The resulting function we denote by X_U^k. To show that this collection of X_U^k does indeed piece together to define a global mapping ${}_kx: M \to G(n,k)$, we must prove the analogue of (3.3), i.e., suppose X_U^k is defined with the help of a coordinate function z and a reduced representation x^* and X_V^* is defined with the help of a coordinate function w and a reduced representation y^*, then on $U \cap V$, we must prove the existence of holomorphic functions h_1, h_2 such that $h_1X_U^k = h_2X_V^k$. This is not difficult: by the remark at the end of Chapter III §1, there exist holomorphic functions g_1, g_2 on $U \cap V$ such that $g_1x^* = g_2y^*$, so a simple computation shows that $g_1^kX_U^k = g_2^k(\frac{dw}{dz})^{\frac{k(k+1)}{2}} X_V^k$. This shows that each associated holomorphic curve of rank k, ${}_kx: M \to G(n,k) \subseteq P_{\ell(k)-1}\mathbb{C}$, is well-defined. The analogue of Lemma 3.6 can be proved in a similar manner, so ${}_kx(M)$ never lies in any polar divisor Σ_B of $G(n,k)$.

Now, according to Theorem 4.2,

$$n_k(M,B) = v_k(M) \tag{4.40}$$

for <u>every</u> $B \in G(n,k)$, where $v_k(M) = \frac{1}{\pi}\int_M {}_kx^*\omega$, and $n_k(M,B)$ = the number of zeroes of $\langle {}_kx, B\rangle$ in M, counting multiplicity. In particular, for any A, $B \in G(n,k)$,

$$n_k(M,A) = n_k(M,B).$$

We proceed to give several interpretations of $v_k(M)$ and $n_k(M,B)$ on the basis of (4.40). First of all, we have remarked previously (above Theorem 4.2) that if $P_1\mathbb{C}$ is any complex line in $P_{\ell(k)-1}\mathbb{C}$, then $\int_{P_1\mathbb{C}} \frac{1}{\pi}\omega = 1$. This shows that the cohomology class $\frac{1}{\pi}\omega$ is in fact the generator of $H^2(P_{\ell(k)-1}\mathbb{C};\mathbb{Z})$. Rewriting $v_k(M)$ as $\int_{{}_kx(M)} (\frac{1}{\pi}\omega)$, we see that $v_k(M)$ is nothing but that integer which, when multiplied with the fundamental two-cycle of $P_{\ell(k)-1}\mathbb{C}$, gives the cycle ${}_kx(M)$. Let now Π_B be the hyperplane $\{\Lambda: \langle\Lambda,B\rangle = 0\}$ in $P_{\ell(k)-1}\mathbb{C}$; Π_B generates the homology group $H_*(P_{\ell(k)-1}\mathbb{C};\mathbb{Z})$ in dimension $2(\ell(k)-1) - 2$, and is hence the Poincaré dual of $\frac{1}{\pi}\omega$. By Poincaré duality and the preceding interpretation of $v_k(M)$, $v_k(M)$ is therefore the intersection number of the cycles ${}_kx(M)$ and Π_B. Since B is arbitrary, we see that:

(4.41) $\quad v_k(M)$ = the order of the algebraic curve $_kx(M)$ in $P_{\ell(k)-1}\mathbb{C}$ in the sense of algebraic geometry.

Now $_kx(M) \subseteq G(n,k)$ and $\Pi_B \cap G(n,k) = \Sigma_B$. So we can also equate $v_k(M)$ with the intersection number of the cycles $_kx(M)$ and Σ_B in $G(n,k)$. In §2 of Chapter I, we pointed out that Σ_B is the generator of $2\{(n-k)(k+1)-1\}$-dimensional integral homology group of $G(n,k)$, so if c_1 is the Poincaré dual of Σ_B, c_1 is the generator of $H^2(G(n,k);\mathbb{Z})$. Let γ_2 be the dual element of c_1 under the natural pairing $H^2 \otimes H_2 \to \mathbb{Z}$, then γ_2 is the generator of $H_2(G(n,k);\mathbb{Z})$, which we call the <u>fundamental two-cycle</u> of $G(n,k)$. Thus Poincaré duality easily leads to:

(4.42) $\quad {}_kx(M) = v_k(M)\cdot\gamma_2$ as homology classes.

Now each $_kx(m)$ $(m \in M)$ is a k-plane in $P_n\mathbb{C}$, so the union $\bigcup_{m\in M} {}_kx(m)$ is a subset of $P_n\mathbb{C}$ which is in fact an algebraic variety (called the <u>variety of osculating k-spaces</u> of the algebraic curve x). We also denote this subvariety of $P_n\mathbb{C}$ by $_kx(M)$. As remarked after (4.4), $n_k(M,B)$ is in fact the intersection number of $_kx(M)$ and the polar space $B^\perp$ of B. $B^\perp$ is a projective subspace of dimension $(n-k-1)$, but is otherwise arbitrary, so (4.40) leads to:

(4.43) $\quad v_k(M)$ = the order of the $(k+1)$-dimensional algebraic variety $_kx(M)$ in $P_n\mathbb{C}$ in the sense of algebraic geometry.

Now, what does Theorem 4.12 have to say about ${}_kx$? Let $\Omega_k = {}_kx^*\omega$ and let K_k be the corresponding Gaussian curvature. Then,

$$(4.44) \qquad 2\pi(\chi(M) + s(M)) = \int_M K_k\Omega_k \ .$$

Take a coordinate neighborhood U of M with coordinate function z; by (4.18) we have:

$$\Omega_k = \frac{\sqrt{-1}}{2} \frac{|X_z^{k-1}|^2 |X_z^{k+1}|^2}{|X_z^k|^4} dz \wedge d\bar{z} \ .$$

Consequently, (2.13) implies that

$$\begin{aligned} K_k\Omega_k &= \frac{-1}{2} dd^c \log\left(\frac{|X_z^{k-1}|^2 |X_z^{k+1}|^2}{|X_z^k|^4}\right) \\ &= -\{dd^c \log |X_z^{k-1}| - 2\, dd^c \log |X_z^k| + dd^c \log |X_z^{k+1}|\} \\ &= -2\{{}_{k-1}x^*\omega - 2{}_kx^*\omega + {}_{k+1}x^*\omega\} \end{aligned}$$

where the last step made use of (4.17). Hence by definition:

$$\int_M K_k\Omega_k = -2\pi\{v_{k-1}(M) - 2v_k(M) + v_{k+1}(M)\}$$

This coupled with (4.44) give us the final result:

$$(4.45) \qquad s(M) + v_{k-1}(M) - 2v_k(M) + v_{k+1}(M) = -\chi(M),$$

for $k = 0,\ldots,n-1$, where $v_{-1}(M) = 0$ by definition and $v_n(M) = 0$ for obvious reasons. (4.45) together with any one of the interpretations of $v_k(M)$ given in (4.41)-(4.43) are known as the <u>Plücker formulas.</u> Their beauty lies in the fact that the left-side of (4.45) consists of analytic invariants of x, while the right-side is a purely topological invariant of M.

§6. We now resume the assumption that V is an open Riemann surface with a harmonic exhaustion τ and proceed to integrate Theorem 4.12. Again recall that we always work in the domain of harmonicity of τ, i.e., in $V - V[r(\tau)]$, so all parameter values are greater than $r(\tau)$.

Let $f: V \to M$ be a nonconstant holomorphic mapping as in Theorem 4.12. If $\partial V[r]$ contains no critical point of f and if r is not a critical value of τ, then in the notation of that theorem,

$$(4.46) \qquad 2\pi\chi(r) + 2\pi s(r) - \int_{\partial V[r]} \kappa = \int_{V[r]} K\Omega$$

where for convenience we have written

$$\chi(r) = \chi(V[r]),$$
$$s(r) = s(V[r]).$$

Let us define a function h in $V - V[r(\tau)]$ by

$$(4.47) \qquad f^*\omega = h \, d\tau \wedge *d\tau$$

Clearly h is not defined at the isolated point set which consists of the critical points of f and τ. The important thing to note is that h <u>is nonnegative</u> because both $\Omega = f^*\omega$ and $d\tau \wedge * d\tau$ are coherent with the naturally given orientation of V. As in the FMT, we convert the line integral of (4.46) into a derivative:

Lemma 4.13. If $\partial V[r]$ does not contain any critical points of f and τ, then

$$\int_{\partial V[r]} \kappa = \frac{d}{dr}\left(\frac{1}{2}\int_{\partial V[r]} (\log h) * d\tau\right).$$

Proof. We use special coordinate function $\sigma = \tau + \sqrt{-1}\rho$ on a component W of $\tau^{-1}((r_1, r_2))$ where $r \in (r_1, r_2)$. Then $\Omega = h\, d\tau \wedge d\rho$ so that $G = f^* F = h(d\tau^2 + d\rho^2)$. By (2.14), $\kappa = \frac{1}{2}\frac{\partial \log h}{\partial \tau} d\rho$. The argument given in Lemma 2.5 now serves to conclude the proof of this lemma also. Q.E.D.

The analogue of Lemma 4.4 is

Lemma 4.14. $\int_{\partial V[r]} (\log h) * d\tau$ is a continuous function of r for all $r \geq r(\tau)$.

*Proof. If $\partial V[r]$ contains no critical point of f and τ, the lemma is trivial. So assume that it contains both. Take a p such that $d\tau(p) = 0$, and $df(p) = 0$. Let $\sigma = \tau + \sqrt{-1}\rho$ be the usual holomorphic function in a neighborhood of p. σ is no longer a coordinate function at p because $d\tau(p) = 0$. Let $\nu = \tau - \tau(p)$ and let ρ be chosen so that $\rho(p) = 0$. Then $\zeta = \nu + \sqrt{-1}\rho$ is a holomorphic function near p such that $\zeta(p) = 0$. For some positive integer m, $z = (\zeta)^{1/m}$ will be a coordinate function in a neighborhood W of p. From the proof of Theorem 4.12, we know that $\Omega = f^*\omega = |z|^{2(\delta-1)}(bA\bar{b}^t)dz \wedge d\bar{z}$, where $bA\bar{b}^t$ is a C^∞ function which vanishes nowhere in W and $(\delta-1)$ is the stationary index of f at p. By definition of h, $\Omega = h\, d\tau \wedge * d\tau$ $= \frac{\sqrt{-1}}{2} h\, d\zeta \wedge d\bar{\zeta}$. Since $d\zeta = mz^{m-1}dz$, we see that

$$\frac{\sqrt{-1}}{2} |z|^{2(\delta-1)}(bA\bar{b}^t)dz \wedge d\bar{z} = \Omega = \frac{\sqrt{-1}}{2} h\, d\zeta \wedge d\zeta$$
$$= \frac{\sqrt{-1}}{2} h \cdot m^2 \cdot |z|^{2(m-1)} dz \wedge d\bar{z}.$$

Hence in W, $h = \frac{1}{m^2} |z|^{2(\delta-m)}(bA\bar{b}^t)$, and so $\log h$ $= 2(\delta-m)\log |z| + C^\infty$ function. Thus on $\partial V[r]$, $\log h$ has at worst only a simple logarithmic singularity. Therefore this lemma may be proved in a manner similar to Lemma 4.4.

Q.E.D.

The situation here parallels that of the FMT. We have by (4.46) and Lemma 4.13 that

$$\chi(r) + s(r) - \frac{d}{dr}\{\frac{1}{4\pi} \int_{\partial V[r]} (\log h) * d\tau\} = \frac{1}{2\pi} \int_{V[r]} K\Omega$$

provided $\partial V[r]$ contains no critical point of f and τ. We integrate this with respect to r and use Lemma 4.14 to conclude that for $r \geq r(\tau)$:

$$\int_{r_0}^{r} \chi(t)dt + \int_{r_0}^{r} s(t)dt - \frac{1}{4\pi} \int_{\partial V[t]} (\log h) * d\tau \Big|_{r_0}^{r}$$
$$= \frac{1}{2\pi} \int_{r_0}^{r} dt \int_{V[t]} K\Omega \,.$$

Introduce the notation:

$$E(r) = \int_{r_0}^{r} \chi(t)dt$$
$$S(r) = \int_{r_0}^{r} s(t)dt$$

and we have:

$$(4.48)\quad E(r) + S(r) - \frac{1}{4\pi}\int_{\partial V[t]}(\log h)*d\tau\Big|_{r_0}^{r} = \frac{1}{2\pi}\int_{r_0}^{r} dt \int_{V[t]} K\Omega.$$

Now recall that for $r \geq r(\tau)$, $\int_{\partial V[r]} *d\tau = L$ is a constant. (Corollary to Lemma 2.4). By the concavity of the logarithm (Lemma 2.14),

$$\frac{L}{4\pi}\log\{\frac{1}{L}\int_{\partial V[r]} h*d\tau\} \geq \frac{1}{4\pi}\int_{\partial V[r]}\log h *d\tau$$
$$= E(r) + S(r) - \frac{1}{2\pi}\int_{r_0}^{r} dt \int_{V[t]} K\Omega + \frac{1}{4\pi}\int_{\partial V[r_0]}(\log h)*d\tau.$$

So if we let $\varphi(r) = \frac{4\pi}{L}\{E(r) + S(r) - \frac{1}{2\pi}\int_{r_0}^{r} dt \int_{V[t]} K\Omega + \text{const.}\}$, then

$$e^{\varphi(r)} \leq \frac{1}{L}\int_{\partial V[r]} h *d\tau,$$

implying, $L\int_{r_0}^{r} e^{\varphi(t)}dt \leq \int_{r_0}^{r} dt \int_{\partial V[t]} h *d\tau$. Now $h\, d\tau \wedge *d\tau = f^*\omega$, which is certainly integrable on $V[r] - V[r_0]$. So Fubini's theorem gives:

$$\int_{r_0}^{r} dt \int_{\partial V[t]} h*d\tau = \int_{V[r]-V[r_0]} h\, d\tau \wedge *d\tau$$
$$\leq \int_{V[r]} h\, d\tau \wedge *d\tau$$
$$= \int_{V[r]} f^*\omega.$$

Combined with the above, this leads to

$$(4.49) \qquad L\int_{r_o}^{r} e^{\varphi(t)}dt \le \int_{V[r]} f^*\omega \,.$$

For this holomorphic map $f\colon V \to M$, let us define $v(r) = \frac{1}{\pi}\int_{V[r]} f^*\omega$,

$$T(r) = \int_{r_o}^{r} v(t)dt.$$

If M is taken to be $G(n,k) \subseteq P_{\ell(k)-1}\mathbb{C}$, this $T(r)$ and $v(r)$ of course coincide with the previously defined $T(r)$ and $v(r)$. Now integrate (4.49) once more, and we obtain:

$$(4.50) \qquad L\int_{r_o}^{r} ds \int_{r_o}^{s} e^{\varphi(t)}dt \le T(r)$$

$$\text{where} \quad \varphi(t) = \frac{4\pi}{L}\{E(t) + S(t) - \frac{1}{2\pi}\int_{r_o}^{t} ds \int_{V[s]} K\Omega + \text{const.}\}.$$

(4.48) and (4.50) constitute the Second Main Theorem. We summarize these in the following.

<u>Theorem 4.15 (SMT)</u>. Let $f\colon V \to M$ be a nonconstant holomorphic mapping, where V is open and admits a harmonic exhaustion and M is complex hermitian with metric F and associated two form ω. Let G denote the hermitian metric f^*F on V minus the critical points of f. Let K be the Gaussian curvature of G and $\Omega = f^*\omega$. Then

$$E(r) + S(r) - \frac{1}{4\pi}\int_{\partial V[t]} (\log h)*d\tau \Big|_{r_o}^{r} = \frac{1}{2\pi}\int_{r_o}^{r} dt \int_{V[t]} K\Omega,$$

where h is defined on $V - V[r(\tau)]$ by $\Omega = h\, d\tau \wedge *d\tau$. Furthermore,

$$L\int_{r_o}^{r} ds \int_{r_o}^{s} e^{\varphi(t)}dt \le T(r)$$

where $\varphi(t) = \frac{4\pi}{L}\{E(t) + S(t) - \frac{1}{2\pi}\int_{r_o}^{t} ds \int_{V[s]} K\Omega + \text{const.}\}$.

We now specialize this theorem to the associated holomorphic curves of rank k of a fixed nondegenerate holomorphic curve $x: V \to P_n\mathbb{C}$. So consider ${}_kx: V \to G(n,k) \subseteq P_{\ell(k)-1}\mathbb{C}$ and <u>replace the</u> M <u>of Theorem 4.15 by</u> $G(n,k)$, f <u>by</u> ${}_kx$. What is h in this case? Operating outside the critical points of τ, we always have $\sigma = \tau + \sqrt{-1}\rho$ as a local coordinate function in $V - V[r]$. We can write equivalently $\Omega = \frac{\sqrt{-1}}{2} h\, d\sigma \wedge d\bar{\sigma}$ On the other hand, we have by (4.18) that outside a discrete point set,

$$\Omega = {}_kx^*\omega = \frac{\sqrt{-1}}{2} \frac{|X_\sigma^{k-1}|^2 |X_\sigma^{k+1}|^2}{|X_\sigma^k|^4} d\sigma \wedge d\bar{\sigma} .$$

Thus for ${}_kx$,

$$h = \frac{|X_\sigma^{k-1}|^2 |X_\sigma^{k+1}|^2}{|X_\sigma^k|^4} \tag{4.51}$$

except on a discrete point set (which is negligible in all matters concerning integration). Next, we claim that for ${}_kx$:

$$\frac{1}{2\pi}\int_{V[t]} K\Omega = -\{v_{k-1}(t) - 2v_k(t) + v_{k+1}(t)\}, \tag{4.52}$$

where $v_k(t) = \frac{1}{\pi}\int_{V[t]} {}_kx^*\omega$ etc. It suffices to prove that almost everywhere $K\Omega = -2\{{}_{k-1}x^*\omega - 2{}_kx^*\omega + {}_{k+1}x^*\omega\}$, and we prove the latter locally. Let U be a coordinate neighborhood with coordinate function z. Then by (4.18), except on a discrete set,

$$\Omega = \frac{\sqrt{-1}}{2} \frac{|X_z^{k-1}|^2 |X_z^{k+1}|^2}{|X_z^k|^4} dz \wedge d\bar{z}$$

By (2.13) we have,

$$K\Omega = \frac{-1}{2} dd^c \log(\frac{|X_z^{k-1}|^2 |X_z^{k+1}|^2}{|X_z^k|^4}).$$

$$= -\{dd^c \log |X_z^{k-1}| - 2dd^c \log |X_z^k| + dd^c \log |X_z^{k+1}|$$

By (4.17), except on a discrete set, ${}_k x^* \omega = \frac{1}{2} dd^c \log |X_z^k|$, etc. so we have the desired result. Combining (4.48), (4.51) and (4.52), we have

$$(4.53) \qquad E(r) + S_k(r) + (T_{k-1}(r) - 2T_k(r) + T_{k+1}(r))$$

$$= \frac{1}{2\pi} \int_{\partial V[t]} \log \frac{|X_\sigma^{k-1}|^2 |X_\sigma^{k+1}|^2}{|X_\sigma^k|^4} * d\tau \Big|_{r_o}^{r}$$

where we have written $S_k(r)$ for the function $\int_{r_o}^{r} s_k(t)dt$, where

$s_k(t)$ = sum of the stationary indices of ${}_k x$ in $V[t]$.

Combining (4.50)-(4.52), we obtain the following: for some positive constant C,

$$(4.54) \qquad \int_{r_o}^{r} ds \int_{r_o}^{s} \exp\{\frac{4\pi}{L} \varphi_k(t)\} dt \leq C\, T_k(r)$$

$$\text{where } \varphi_k(t) = \{E(t) + S_k(t) + (T_{k-1}(t) - 2T_k(t) + T_{k+1}(t))\}$$

(4.53) and (4.54) essentially constitute the SMT of rank k, but further refinements of (4.54) are possible and we devote the next section to these.

§7. We begin by introducing a notation first suggested by H. and J. Weyl. Let θ and Φ be continuous functions defined on $[0,s)$, s as in Definition 2.1. We say $\theta = \mu(\Phi)$ if and only if

$$\int_{r_0}^{r} ds \int_{r_0}^{s} \exp\{K\theta(t)\} dt < C\Phi(r) + C'$$

for some positive constants K, C and C'. (Actually the Weyl's wrote $\theta = \omega(\Phi)$, but we have already used ω to denote the Kahler form of the F-S metric on $P_n\mathbb{C})$. Here are the basic properties:

Lemma 4.16.

(i) If $\theta_1 \leq \theta$ off a compact set and $\Phi \leq \Phi_1$ off a compact set, then $\theta = \mu(\Phi)$ implies $\theta_1 = \mu(\Phi_1)$.

(ii) If $\theta = \mu(\Phi)$, then $\theta + O(1) = \mu(\Phi)$.

(iii) If C is a positive constant and $\theta = \mu(\Phi)$, then $C\theta = \mu(\Phi)$.

(iv) If $\theta = \mu(\Phi)$ and θ_1 is positive off a compact set, then $\theta - \theta_1 = \mu(\Phi)$.

(v) Suppose $\theta = \mu(\Phi)$ and $\theta_1 = \mu(\Phi)$, then $\theta + \theta_1 = \mu(\Phi)$.

Proof. (i)-(iv) are obvious. For (v), use Schwarz's inequality twice. Q.E.D.

When $\theta = \mu(\Phi)$, we have immediately a comparison between the growths of θ and $\log \Phi$ as $r \to s$. See Lemmas 2.16 and 2.19. This will be made precise later. For the moment, let us rewrite (4.54) as:

(4.55) $$E + S_k + (T_{k-1} - 2T_k + T_{k+1}) = \mu(T_k)$$

for $k = 0,\ldots,n-1$. It may be recalled that $T_{-1}(r) \equiv 0$ as a consequence of the definition $X^{-1} \equiv 1$, and $T_n(r) \equiv 0$ because $dd^c \log |X^n| \equiv 0$ (for X^n is nothing but $\psi\epsilon_0 \wedge \cdots \wedge \epsilon_n$, where ψ is a holomorphic function, so $|X^n| = |\psi|$). Now S_k is given by the stationary index of ${}_kx$, so by definition $S_k \geq 0$. We may therefore use Lemma 4.16 (iv) to simplify (4.55) to:

(4.56) $$E + (T_{k-1} - 2T_k + T_{k+1}) = \mu(T_k)$$

valid for $k = 0,\ldots,n-1$. We will now show that this inequality leads to the remarkable fact that the various order functions $T_0,\ldots,T_{n-1}$ are essentially of the same order of magnitude. To this end, we define a new function $T\colon [0,s) \to \mathbb{R}$ by: for each $r \in [0,s)$;

$$T(r) = \max\{T_0(r),\ldots,T_{n-1}(r)\}.$$

By Lemma 4.16(i), (4.55) and (4.56) imply:

(4.57) $$E + S_k + (T_{k-1} - 2T_k + T_{k+1}) = \mu(T),$$

(4.58) $$E + (T_{k-1} - 2T_k + T_{k+1}) = \mu(T),$$

for $k = 0,\ldots,n-1$.

Lemma 4.17. For $k = 0,\ldots,n-2$,

$(4.59)_k$ $$(k+1)T_{k+1} = (k+2)T_k - \frac{(k+2)(k+1)}{2} E + \mu(T),$$

and for $k = 1,\dots,n-1$,

$$(4.60)_k \quad (n-k)T_{k-1} = (n-k+1)T_k - \frac{(n-k)(n-k+1)}{2} E + \mu(T).$$

Proof. We first prove (4.59). It is true for $k = 0$ because in this case it reads: $T_1 = 2T_0 - E + \mu(T)$, which is (4.58) for $k = 0$. Now assume $(4.59)_{k-1}$, i.e., assume

$$kT_k = (k+1)T_{k-1} - \frac{k(k+1)}{2} E + \mu(T)$$

for $k > 0$. We will prove $(4.59)_k$ by induction. Rewrite the above as $\frac{k(k+1)}{2} E - (k+1)T_{k-1} + kT_k = \mu(T)$. By Lemma 4.16(iii), $(k+1)$ times (4.58) gives $(k+1)E + (k+1)T_{k-1} - 2(k+1)T_k + (k+1)T_{k+1} = \mu(T)$. Adding these two and using Lemma 4.16(v) we obtain exactly $(4.59)_k$.

To prove (4.60), we use backward induction. The case of $k = n-1$ reads: $T_{n-2} = 2T_{n-1} - E + \mu(T)$. But this is just a restatement of (4.58) for $k = n - 1$. (Recall $T_n \equiv 0$).
Now assume $(4.60)_{k+1}$, i.e., assume

$$(n-k-1)T_k = (n-k)T_{k+1} - \frac{(n-k-1)(n-k)}{2} E + \mu(T)$$

Rewrite this as $\frac{(n-k-1)(n-k)}{2} E + (n-k-1)T_k - (n-k)T_{k+1} = \mu(T)$. Multiply (4.58) by $(n-k)$ and use Lemma 4.16(iii) to obtain: $(n-k)E + (n-k)T_{k-1} - 2(n-k)T_k + (n-k)T_{k+1} = \mu(T)$. Add these two and use Lemma 4.16(v), we get exactly $(4.60)_k$. Q.E.D.

First Corollary. If $\ell \geq k$, then

$$(k+1)T_\ell = (\ell+1)T_k - \frac{(\ell-k)}{2}(k+1)(\ell+1)E + \mu(T)$$

If $\ell \leq k$, then

$$(n-k)T_\ell = (n-\ell)T_k - \frac{(k-\ell)}{2}(n-\ell)(n-k)E + \mu(T).$$

Proof. Straightforward induction from the lemma.

Second Corollary. If $V = \mathbb{C}$, or $\mathbb{C} - \{0\}$, or a disc, or an annulus, then for $\ell \geq k$,

$$(k+1)T_\ell = (\ell+1)T_k + \mu(T),$$

and for $\ell \leq k$,

$$(n-k)T_\ell = (n-\ell)T_k + \mu(T).$$

Proof. In any of these cases, $\chi(V) \geq 0$, so for sufficiently large r, $\chi(r) \geq 0$ (see discussion after Theorem 2.17). Thus $E(r) \geq 0$ for r large. The desired result now follows from the First Corollary and Lemma 4.16(iv). Q.E.D.

Lemma 4.18. For $k = 0,\dots,n-1$

$$S_k = \frac{(n+1)}{(k+1)(n-k)} T_k - \frac{(n+1)}{2}E + \mu(T) \tag{4.61}$$

*Proof. For $k = 0$, (4.57) gives $S_0 = 2T_0 - T_1 - E + \mu(T)$, while $(4.60)_1$ gives by virtue of Lemma 4.16(iii) that $-T_1 = -\frac{(n-1)}{n}T_0 - \frac{(n-1)}{2}E + \mu(T)$. These two together give $S_0 = \frac{(n+1)}{n}T_0 - \frac{(n+1)}{2}E + \mu(T)$, using Lemma 4.16(v) in the process. So (4.61) is proved for $k = 0$. The case of $k = n-1$ is proved in a similar manner. Now let $1 \leq k \leq n-2$. $(4.59)_{k-1}$ gives

$$- T_{k-1} = - \frac{k}{(k+1)} T_k - \frac{k}{2} E + \mu(T),$$

while $(4.60)_{k+1}$ gives

$$- T_{k+1} = - \frac{(n-k-1)}{(n-k)} T_k - \frac{(n-k-1)}{2} E + \mu(T)$$

(We have used Lemma 4.16(iii) twice). Now substitute these into (4.57) and use Lemma 4.16(v) to simplify, we get (4.61).

Q.E.D.

Corollary. If V is $\mathbb{C}$, $\mathbb{C} - \{0\}$, a disc or an annulus, then for $k = 0,\ldots,n-1$,

$$S_k = \frac{(n+1)}{(k+1)(n-k)} T_k + \mu(T)$$

Proof. Same reasoning as the Second Corollary of the preceding lemma.

At this point, we must distinguish between the cases of infinite and finite harmonic exhaustions. (Definition 2.4 of §6 of Chapter II). Leaving the finite case to the reader as an exercise (cf. the end of Chapter II), we concentrate on the infinite case. Thus let $\tau\colon V \to [0,\infty)$. Recall that $\theta = \mu(\Phi)$ if and only if

$$\int_{r_0}^{r} ds \int_{r_0}^{s} \exp\{K\theta(t)\}dt < C\Phi(r) + C'$$

for some positive constants K, C and C'. By applying Lemma 2.16 twice we see that $\theta = \mu(\Phi)$ implies

$$(4.62) \qquad \| \ \theta(r) < \kappa \log(C\Phi(r) + C')$$

where as in §6 of Chapter II, $\kappa > 1$, and the sign "$\|$" in front of an inequality means that the inequality is only valid in $[0,\infty) - I$ with $\int_I d \log x < \infty$. Applying (4.62) to the Second Corollary of Lemma 4.17, we get: <u>if</u> V <u>is</u> $\mathbb{C}$ <u>or</u> $\mathbb{C} - \{0\}$, then

$$\| \ (k+1)T_\ell(r) < (\ell+1)T_k(r) + \kappa \log(CT(r) + C') \quad \text{for} \quad \ell \geq k,$$

$$\| \ (n-k)T_\ell(r) < (n-\ell)T_k(r) + \kappa \log(CT(r) + C') \quad \text{for} \quad \ell \leq k.$$

Obviously, they together imply that for some positive constants c_1 and c_2,

$$\| \ T(r) < c_1 T_k(r) + c_2 \log(CT(r) + C')$$

Now by definition, $T_k(r) = \int_{r_0}^r dt \int_{V[t]} {}_k x^* \omega$, so $T_k(r) \to \infty$ as $r \to \infty$ because $\int_{V[t]} {}_k x^* \omega$ as a function of t is strictly increasing. A fortiori, $T(r) \to \infty$ as $r \to \infty$. Consequently, $\lim_{r\to\infty} \frac{\log T(r)}{T(r)} = 0$. Thus for sufficiently large r, $c_2 \log(CT(r) + C') < \frac{1}{2} T(r)$. Combined with the inequality above, we have that for sufficiently large r,

$$\| \ T(r) < 2c_1 T_k(r)$$

for $k = 0,\ldots,n-1$. Note that since "$\|$" indicates inequality only valid outside an exception set I on which $\int_I d \log x < \infty$, we may as well incorporate any compact interval into I. Therefore, the preceding inequality is valid without the redundant phrase: "for sufficiently large r". So we can say that there exists a positive constant e such that

(4.63) $$\| \ T(r) < eT_k(r)$$

for all $k = 0,\ldots,n-1$.

Next, we apply (4.62) to the Corollary of Lemma 4.18: if $V = \mathbb{C}$ or $\mathbb{C} - \{0\}$,

$$\| \ S_k(r) < \frac{(n+1)}{(k+1)(n-k)} T_k(r) + \kappa \log(CT(r) + C')$$

By (4.63), the last term may be replaced by $\kappa \log(CeT_k(r) + C')$. But since $\lim_{r\to\infty} \frac{\log T_k(r)}{T_k(r)} = 0$, given any ϵ, the inequality $\kappa \log(CeT_k(r) + C') < \epsilon \, T_k(r)$ would hold outside a sufficiently large compact set. This implies

(4.64) $$\| \ S_k(r) < \left(\frac{(n+1)}{(k+1)(n-k)} + \epsilon\right)T_k(r).$$

We may summarize (4.63) and (4.64) into

Lemma 4.19. If V is $\mathbb{C}$ or $\mathbb{C} - \{0\}$, then there exists a positive constant e such that for all $k = 0,\ldots,n-1$,

$$\| \ T_k(r) \leq T(r) < eT_k(r).$$

Moreover, given any $\epsilon > 0$,

$$\| \ S_k(r) < \left(\frac{(n+1)}{(k+1)(n-k)} + \epsilon\right)T_k(r).$$

In this sense all $T_0,\ldots,T_{n-1}$ are of the same order of magnitude, and S_k does not grow faster than a fixed multiple of T_k. As an application of this, we prove

Lemma 4.20. Let V be either $\mathbb{C}$ or $\mathbb{C} - \{0\}$. Then it is impossible to have $T_k + CE = \mu(T^2)$ for any positive constant

C and for any $k = 0,\ldots,n-1$.

Proof. We noted previously that $E(r)$ is nonnegative for sufficiently large r in case $V = \mathbb{C}$ or $\mathbb{C} - \{0\}$, so by Lemma 4.16(iv), the hypothesis implies that $T_k = \mu(T^2)$. By (4.62) this entails

$$\| \; T_k(r) < \kappa \log(CT^2(r) + C')$$

Lemma 4.19 implies that,

$$\begin{aligned} \| \; \kappa \log(CT^2(r) + C') &< \kappa \log(Ce^2T_k^2(r) + C') \\ &< 2\kappa \log(T_k^2(r)) \\ &= 4\kappa \log T_k(r). \end{aligned}$$

Thus $\| \; T_k(r) < 4\kappa \log T_k(r)$, and so

$$1 \leq \limsup_{r\to\infty} \frac{4\kappa \log T_k(r)}{T_k(r)} = 0,$$

a contradiction. Q.E.D.

Now suppose $V \neq \mathbb{C}$ or $\mathbb{C} - \{0\}$, (but we still assume that V has an infinite harmonic exhaustion.) We cannot expect Lemmas 4.19 and 4.20 to hold without further restrictions because in this case, $E < 0$. Motivated by the definition of transcendency in Chapter II, we are led to imposing the same condition on holomorphic curves.

Definition 4.1. A holomorphic curve $x\colon V \to P_n\mathbb{C}$ is called transcendental if and only if $\lim_{r\to\infty} \frac{r}{T_0(r)} = 0$.

We claim that if $\chi(V)$ is finite and x is transcendental, then $\lim_{r\to\infty} \frac{E(r)}{T_0(r)} = 0$. The proof is the same as in Chapter II: let r_0 be such that $r \geq r_0$ implies $\chi(r) = \chi(V)$, then

$$\lim_{r\to\infty} \frac{E(r)}{T_0(r)} = \lim_{r\to\infty} \frac{\int_{r_0}^{r} \chi(t)\,dt}{T_0(r)} = \lim_{r\to\infty} \frac{\chi(r_0)(r-r_0)}{T_0(r)} = 0.$$

Hence assuming $\chi(V)$ finite and x transcendental, if $V \neq \mathbb{C}$ or $\mathbb{C} - \{0\}$, we have for all sufficiently large r that

$$-\frac{(\ell-k)}{2}(k+1)(\ell+1)E(r) \leq \epsilon T_0(r) \leq \epsilon T(r)$$

for $0 \leq k \leq \ell \leq n-1$ and

$$-\frac{(k-\ell)}{2}(n-\ell)(n-k)E(r) \leq \epsilon T_0(r) \leq \epsilon T(r)$$

for $0 \leq \ell \leq k \leq n-1$, where ϵ is any preassigned positive number. Therefore according to the First Corollary of Lemma 4.17 and (4.62),

$$\| \quad (k+1)T_\ell(r) \leq (\ell+1)T_k(r) + (\epsilon T(r) + \kappa \log(CT(r) + C'))$$

if $k \leq \ell$ and

$$\| \quad (n-k)T_\ell(r) \leq (n-\ell)T_k(r) + (\epsilon T(r) + \kappa \log(CT(r) + C'))$$

if $\ell \leq k$. Together these two imply the existence of positive constants c_1 and c_2 (independent of ϵ) such that

$$\| \quad T(r) \leq c_1 T_k(r) + c_2(\epsilon T(r) + \kappa \log(CT(r) + C')).$$

Then again, $T(r) \to \infty$ as $r \to \infty$ implies $\lim_{r\to\infty} \frac{\log T(r)}{T(r)} = 0$, implies $\kappa \log(CT(r) + C') \leq \epsilon T(r)$ for all sufficiently large r. Hence,

$$\| \; T(r) \leq c_1 T_k(r) + 2\epsilon c_2 T(r)$$

If we choose ϵ so small that $2\epsilon c_2 < \frac{1}{2}$ (we may do this because c_2 is independent of ϵ) we have:

$$\| \; T(r) \leq 2c_1 T_k(r) \tag{4.65}$$

for some positive constant c_1, for all $k = 0,\ldots,n-1$.

Still assuming the transcendency of x and the finiteness of $\chi(V)$, we next apply (4.62) to Lemma 4.18. In a similar way to the above, we see that for any preassigned $\epsilon > 0$,

$$\begin{aligned} \| \; S_k(r) &\leq \frac{(n+1)}{(k+1)(n-k)} T_k(r) + \epsilon T(r) + \kappa \log(CT(r) + C') \\ &\leq \frac{(n+1)}{(k+1)(n-k)} T_k(r) + 2\epsilon T(r) \\ &\leq \frac{(n+1)}{(k+1)(n-k)} T_k(r) + 4\epsilon c_1 T_k(r). \end{aligned}$$

Since ϵ is arbitrary and c_1 is independent of ϵ, we see that

$$\| \; S_k(r) \leq \left(\frac{(n+1)}{(k+1)(n-k)} + \epsilon\right) T_k(r) \tag{4.66}$$

for any prescribed $\epsilon > 0$. We summarize (4.65) and (4.66) in

Lemma 4.21. Suppose $V \neq \mathbb{C}$ or $\mathbb{C} - \{0\}$, $\chi(V)$ is finite and V has an infinite harmonic exhaustion. If $x: V \to P_n\mathbb{C}$ is transcendental, then there exists a positive constant e such that

$$\| \quad T_k(r) \leq T(r) \leq eT_k(r)$$

for all $k = 0,\ldots,n-1$. Furthermore, given any $\epsilon > 0$,

$$\| \quad S_k(r) \leq \left(\frac{(n+1)}{(k+1)(n-k)} + \epsilon\right)T_k(r).$$

<u>Corollary</u>. If assumptions as above, then $\liminf_{r\to\infty} \frac{r}{T_k(r)} = 0$ for all $k = 1,\ldots,n-1$.

<u>Proof</u>. We have

$$\| \quad \frac{1}{T_k(r)} \leq e \cdot \frac{1}{T(r)} \leq e \cdot \frac{1}{T_0(r)},$$

while $\limsup_{r\to\infty} \frac{r}{T_0(r)} = 0$ by assumption of transcendency of x. So the conclusion is immediate. Q.E.D.

<u>Lemma 4.22</u>. Suppose $V = \mathbb{C}$ or $\mathbb{C} - \{0\}$, $\chi(V)$ is finite and V has an infinite harmonic exhaustion. If $x: V \to P_n\mathbb{C}$ is transcendental, then it is impossible to have $T_k + c_1E = \mu(T^2)$ for any constant c_1 and for any $k = 0,\ldots,n-1$

<u>Proof</u>. By (4.62), we have

$$\| \quad T_k(r) + c_1E(r) \leq \kappa \log(CT^2(r) + C')$$

By Lemma 4.21, there is a positive constant e such that

$$\| \quad \frac{1}{e} T(r) + c_1E(r) \leq \kappa \log(CT^2(r) + C')$$

For sufficiently large r, we know that $\kappa \log(CT^2(r) + C') \leq 2\kappa \log T^2(r) = 4\kappa \log T(r)$. Hence,

$$\| \quad \frac{1}{e} T(r) + c_1E(r) \leq 4\kappa \log T(r)$$

or, $\frac{1}{e} + \liminf_{r\to\infty} c_1 \frac{E(r)}{T(r)} \leq \limsup_{r\to\infty} 4\kappa \frac{\log T(r)}{T(r)}$. But $T_0(r) \leq T(r)$ and we have seen above that $\lim_{r\to\infty} \frac{E(r)}{T_0(r)} = 0$ if x is transcendental, so

$$\liminf_{r\to\infty} c_1 \frac{E(r)}{T(r)} = \lim_{r\to\infty} c_1 \frac{E(r)}{T(r)} = 0 .$$

The lim sup on the right is of course zero. We are therefore left with $\frac{1}{e} \leq 0$, a contradiction. Q.E.D.

Finally, we explain the meaning of transcendency by proving the analogue of Lemma 2.18 in Chapter II.

Lemma 4.23. Let $V = M - \{a_1,\ldots,a_m\}$, where M is a compact Riemann surface, and $a_i \in M$, $i = 1,\ldots,m$. Then $x\colon V \to P_n\mathbb{C}$ is transcendental if and only if x cannot be extended to a holomorphic mapping $x'\colon M \to P_n\mathbb{C}$.

Proof. We first prove that if x is transcendental, then it is not extendable to x'. Suppose it is extendable to x', then $\lim_{r\to\infty} x(V - V[r]) = \{x'(a_1),\ldots,x'(a_m)\}$, and so there will be at least one hyperplane Π of $P_n\mathbb{C}$ with the following property: there exists a neighborhood U of Π such that $U \cap x(V - V[r]) = \emptyset$ for all sufficiently large r. Let $a \in P_n\mathbb{C}$ be the point orthogonal to Π. The existence of such a U clearly implies that the restriction of $|\langle x,a\rangle|$ to $V - V[r]$ is bounded below for all large r, and hence the restriction of $\log \frac{|\tilde{x}|}{|x,a|}$ to $V - V[r]$ is bounded above by K for all large r. Let us assume that r_0 is sufficiently large. Then obviously $n(t,a) = n(r_0,a)$ for all $t \geq r_0$, and so

$$N(r,a) = \int_{r_0}^{r} n(t,a)dt = n(r_0,a)(r-r_0).$$

Now we use (4.16) and (4.10) to obtain:

$$\begin{aligned} T_0(r) &= N_0(r,a) + \frac{1}{2\pi}\int_{\partial V[t]} \log \frac{|\tilde{x}|}{|\tilde{x},a|} *d\tau \Big|_{r_0}^{r} \\ &\leq N_0(r,a) + \frac{1}{2\pi}\int_{\partial V[r]} \log \frac{|\tilde{x}|}{|\tilde{x},a|} *d\tau \\ &\leq n(r_0,a)(r-r_0) + \frac{K}{2\pi}\cdot L \end{aligned}$$

where $L = \int_{\partial V[r]} *d\tau$ as usual. So $\limsup_{r\to\infty} \frac{r}{T(r)} \geq \frac{1}{n(r_0,a)} > 0$. This contradicts transcendency.

To prove the converse, let us assume that x is not transcendental, or more precisely that $\limsup_{r\to\infty} \frac{r}{T_0(r)} = \beta > 0$, and we will show how to extend x to $x': M \to P_n\mathbb{C}$. We begin with the observation that if we have $(n+1)$ meromorphic functions $\{f_0,\ldots,f_n\}$ defined on a compact (or for that matter, arbitrary) Riemann surface M, then the function $m \mapsto (f_0(m),\ldots,f_n(m))$ induces in a natural way a holomorphic mapping of M into $P_n\mathbb{C}$. For if we let M' be the complement of the finite set of the poles of $\{f_i\}$ and the common zeroes of $\{f_i\}$, we have a mapping $\tilde{f}: M' \to \mathbb{C}^{n+1} - \{0\}$. Define $f = \pi \circ \tilde{f}$, where $\pi: \mathbb{C}^{n+1} - \{0\} \to P_n\mathbb{C}$ is the usual fibration. It suffices to extend f to all of M. We first extend f over the common zeroes of the $\{f_i\}$. This can be done in exactly the same way as in the proof of Lemma 3.1. Then we extend f over any of the poles of the $\{f_i\}$ by treating the poles as we did the zeroes. There is no need to write down the details.

Back to the proof. By (4.25), $N_0(r,a) < T_0(r) + c_0$ for all $a \in P_n\mathbb{C}$, so by our assumption above, we have $\liminf\limits_{r\to\infty} \frac{N_0(r,a)}{r} < \frac{1}{\beta} < \infty$. But l'Hôpital's rule implies that in fact:

$$\liminf_{r\to\infty} \frac{N_0(r,a)}{r} = \lim_{r\to\infty} n_0(r,a)$$

so $\lim\limits_{r\to\infty} n_0(r,a) < \frac{1}{\beta} < \infty$. (See the proof of Lemma 2.18). By the very definition of $n_0(r,a)$, we see that $x(V)$ <u>intersects every hyperplane only a finite number of times</u>, and this intersection number is bound above by $\frac{1}{\beta}$. In particular, $x(V)$ intersects the hyperplane $\{z_0 = 0\}$ only finitely many times. So if $\tilde{x} = (x_0,\ldots,x_n)$ is the mapping into $\mathbb{C}^{n+1}$ which induces x, x_0 has only a finite number of zeroes $\{p_1,\ldots,p_\ell\}$ in V.

Let $U_0 = \{[z_0,\ldots,z_n]\colon z_0 \neq 0\}$, and let $\zeta\colon U_0 \to \mathbb{C}^n$ be the usual coordinate mapping such that $\zeta([z_0,\ldots,z_n]) = ({}^{z_1}\!/z_0,\ldots,{}^{z_n}\!/z_0)$. Consider $\zeta\circ\tilde{x}\colon M - \{a_1,\ldots,a_m\} - \{p_1,\ldots,p_\ell\} \to \mathbb{C}^n$. Then clearly $\zeta\circ\tilde{x} = (\frac{x_1}{x_0},\ldots,\frac{x_n}{x_0})$. Of course, $\zeta\circ\tilde{x}$ has poles at $\{p_1,\ldots,p_\ell\}$. We will show presently that none of the $\{{}^{x_j}\!/x_0\}$ can have an essential singularity at $\{a_1,\ldots,a_m\}$, so ${}^{x_1}\!/x_0,\ldots,{}^{x_n}\!/x_0$ are meromorphic functions on M. By the remarks above, the $(n+1)$ meromorphic functions $(1,\frac{x_1}{x_0},\ldots,\frac{x_n}{x_0})$ induce a holomorphic mapping of M into $P_n\mathbb{C}$, and this mapping obviously extends $x\colon V \to P_n\mathbb{C}$.

It remains to show that $\frac{x_1}{x_0},\ldots,\frac{x_n}{x_0}$ have no essential singularities on M. Suppose $\frac{x_1}{x_0}$ has an essential singularity at a_1. By the Casorati-Weirstrass theorem and the Baire category theorem, there is a dense set in $\mathbb{C}$ each of which has an infinite pre-image. So let $\frac{x_1}{x_0}$ assume a certain value λ an infinite number of times. But this would mean that $x(V)$ intersects the hyperplane defined by $\{z_1 = \lambda z_0\}$ an infinite number of times, contradicting a conclusion reached above. So $\frac{x_1}{x_0},\ldots,\frac{x_n}{x_0}$ have no essential singularities on M. Q.E.D.

We now summarize the main conclusions of §6 and §7 in the following.

Theorem 4.24 (SMT of rank k). Let $x\colon V \to P_n\mathbb{C}$ be a nondegenerate holomorphic curve and let ${}_kx\colon V \to G(n,k) \subseteq P_{\ell(k)-1}\mathbb{C}$ be its associated curve of rank k, $k = 0,\ldots,n-1$. Suppose V admits a harmonic exhaustion, then for each k,

$$E(r) + S_k(r) + (T_{k-1}(r) - 2T_k(r) + T_{k+1}(r))$$
$$= \frac{1}{2\pi}\int_{\partial V[t]} \log \frac{|X_\sigma^{k-1}|^2|X_\sigma^{k+1}|^2}{|X_\sigma^k|^4} *d\tau\Big|_{r_0}^{r}$$

and if $T(r) = \max\{T_0(r),\ldots,T_{n-1}(r)\}$, then

$$E + S_k + (T_{k-1} - 2T_k + T_{k+1}) = \mu(T).$$

$(T_{-1} = T_n \equiv 0)$ Furthermore, assume that either (a) $V = \mathbb{C}$ or $\mathbb{C} - \{0\}$ or (b) V has an infinite harmonic exhaustion, $V \neq \mathbb{C}$ or $\mathbb{C} - \{0\}$, $\chi(V)$ is finite and x is transcendental.

Then there exists a positive constant e such that for all $k = 0,\dots,n-1$

$$\| \quad T_k(r) \leq T(r) \leq eT_k(r).$$

If ϵ is any prescribed positive number, the following inequality also holds:

$$\| \quad S_k(r) \leq \left(\frac{(n+1)}{(k+1)(n-k)} + \epsilon\right)T_k(r)$$

for $k = 0,\dots,n-1$. Finally, under assumptions (a) or (b), it is impossible to have an inequality: $T_k + CE = \mu(T^2)$, where $0 \leq k \leq n-1$ and C is any positive constant. Under assumption (b), this is impossible even for nonpositive C's.

CHAPTER V

The defect relations

§1. We now have both the FMT and the SMT. Following the pattern established in Chapter II, the next move will be to integrate (4.25) for $k = 0$ $(N_0(r,a) < \dot{T}_0(r) + c_0)$ with respect to a function β over $P_n\mathbb{C}$. While it is possible to integrate directly over $P_n\mathbb{C}$, for technical reasons, it will be simpler to lift the domain of integration to $\mathbb{C}^{n+1}$. To this end, we will apply the following lemma to the principal fibration π_2: $S^{2n+1} \to P_n\mathbb{C}$ of Chapter I.

Lemma 5.1. Let π: $P \to M$ be a principal fibre bundle with G as structure group such that (i) G, P and M have riemannian metrics and G acts as a group of isometries, (ii) the map $g \mapsto p \cdot g$ of G into P is an isometric imbedding for each $p \in P$ and (iii) if H denotes the distribution in P which is the orthogonal complement of the tangent space to the fibre, then $d\pi|H$ is an isometry at each point of P. Let Ω_P, Ω_M and Ω_G denote the volume forms of these manifolds and let f be an integrable function on M. Then

$$\left(\int_G \Omega_G\right) \cdot \left(\int_M f\Omega_M\right) = \int_P (\pi^* f)\Omega_P$$

Proof. The proof is trivial because if all these conditions are met, then for every trivializing neighborhood U of M, $\pi^{-1}(U)$ is isometric to $G \times U$ equipped with the product metric. So Fubini's theorem clearly says that

$$\left(\int_G \Omega_G\right) \cdot \left(\int_U f\Omega_M\right) = \int_{\pi^{-1}(U)} (\pi^* f)\Omega_p .$$

A partition of unity then does the rest. Q.E.D.

Now the Corollary to Lemma 1.1 says that the principal fibration $\pi_2: S^{2n+1} \to P_n\mathbb{C}$ (S^{2n+1} and the circle equipped with the usual metrics and $P_n\mathbb{C}$ with the F-S metric) satisfies all three conditions of Lemma 5.1. Hence if β is an integrable function on $P_n\mathbb{C}$,

$$(5.1) \qquad \int_{S^{2n+1}} \pi_2^*\beta\ \Omega_S = 2\pi \int_{P_n\mathbb{C}} \beta\Omega$$

where Ω denotes the volume form of the F-S metric of $P_n\mathbb{C}$, Ω_S denotes the volume form of S^{2n+1} and the volume of the circle is obviously 2π. Again from Chapter I, we have a second principal fibering $\pi_1: \mathbb{C}^{n+1} - \{0\} \to S^{2n+1}$. Let S_r be the sphere of radius r imbedded in $\mathbb{C}^{n+1}$ (thus $S_1 = S^{2n+1}$) and let $\Omega(r)$ be the volume form of S_r in the induced metric (thus $\Omega_S = \Omega(1)$). Let dL denote the usual Lebesgue volume form of $\mathbb{C}^{n+1}$, i.e.,

$$(5.2) \qquad dL \equiv \left(\frac{\sqrt{-1}}{2}\right)^{n+1}(dz_0 \wedge d\bar{z}_0) \wedge \cdots \wedge (dz_n \wedge d\bar{z}_n).$$

Note that if $r = (\Sigma_A\ z_A\bar{z}_A)^{1/2}$, then

$$dL = dr \wedge \Omega(r),$$

so that in terms of the usual fibration $\pi: \mathbb{C}^{n+1} - \{0\} \to P_n\mathbb{C}$ (see (1.1)), we have:

$$\int_{\mathbb{C}^n} \pi^*\beta e^{-\langle Z,Z\rangle} dL = \int_{\mathbb{C}^{n+1}-\{0\}} \pi_1^*(\pi_2^*\beta) e^{-\langle Z,Z\rangle} dL$$
$$= \int_0^\infty e^{-r^2} dr \int_{S_r} \pi_1^*(\pi_2^*\beta)\Omega(r).$$

Now $\pi_1 | S_r$ is a diffeomorphism of S_r onto S^{2n+1} and it is quite clear that $\pi_1^*\Omega_S = \frac{1}{r^{2n+1}}\Omega(r)$, i.e., $\Omega(r) = r^{2n+1}\pi_1^*\Omega_S$. Thus,

$$\int_{\mathbb{C}^n} \pi^*\beta e^{-\langle Z,Z\rangle} dL = \int_0^\infty e^{-r^2} dr \int_{S_r} r^{2n+1}\pi_1^* (\pi_2^*\beta \cdot \Omega_S)$$
$$= \int_0^\infty r^{2n+1} e^{-r^2} dr \int_{S_r} \pi_1^*(\pi_2^*\beta \cdot \Omega_S)$$
$$= \int_0^\infty r^{2n+1} e^{-r^2} dr \int_{S^{2n+1}} (\pi_2^*\beta)\Omega_S .$$
$$= \left(\int_0^\infty r^{2n+1} e^{-r^2} dr\right) \cdot 2\pi \cdot \int_{P_n\mathbb{C}} \beta\Omega \quad \text{(by (5.1))}$$
$$= \frac{1}{2}\int_0^\infty t^n e^{-t} dt \cdot 2\pi \int_{P_n\mathbb{C}} \beta\Omega$$
$$= \frac{1}{2}\Gamma(n+1) \cdot 2\pi \int_{P_n\mathbb{C}} \beta\Omega$$
$$= n!\pi \int_{P_n\mathbb{C}} \beta\Omega.$$

i.e., $\int_{P_n\mathbb{C}} \beta\Omega = \frac{1}{n!\pi} \int_{\mathbb{C}^{n+1}} (\pi^*\beta) e^{-\langle Z,Z\rangle} dL$. We summarize this into the following theorem.

Theorem 5.2. Let β be an integrable function on $P_n\mathbb{C}$ and let Ω and dL denote respectively the volume form of the Fubini-Study metric on $P_n\mathbb{C}$ and the flat metric on $\mathbb{C}^{n+1}$ respectively. Then

$$\int_{P_n\mathbb{C}} \beta\Omega = \frac{1}{n!\pi}\int_{\mathbb{C}^{n+1}} (\pi^*\beta)e^{-\langle Z,Z\rangle}dL$$

<u>Corollary</u>. The volume of $P_n\mathbb{C}$ in the F-S metric is $\pi^n/n!$.

Before we can integrate $N_0(r,a) < T_0(r) + c_0$, we need one more bit of preparatory material.

<u>Lemma 5.3</u>. $n_0(t,a)$ is a measurable function on $[0,s) \times P_n\mathbb{C}$

*<u>Proof</u>. It is obvious that a semi-continuous function is measurable, so we need only prove that $n_0(t,a)$ is semi-continuous. Recall that we have a holomorphic and nondegenerate $x: V \to P_n\mathbb{C}$. For $a \in P_n\mathbb{C}$, $n_0(t,a)$ is by definition the sum of the orders of zeroes of $\langle x,a\rangle$ in $V[t]$. The zeroes are isolated because x is nondegenerate. We will prove $\limsup_{n\to\infty} n_0(t_n,a_n) \leq n_0(t,a)$ for $(t_n,a_n) \to (t,a)$, and this is equivalent to semi-continuity. Let $\{t_n\} = \{r_i\} \cup \{s_j\}$, where $r_i \geq t$, and $s_j \leq t$. It suffices to prove:

$$\limsup_{i\to\infty} n_0(r_i,a_i) \leq n_0(t,a) \tag{5.3}$$

$$\limsup_{j\to\infty} n_0(s_j,a_j) \leq n_0(t,a) \tag{5.4}$$

We will first prove (5.4). In the following, we can localize our considerations to a neighborhood of each of the zeroes of $\langle x,a\rangle$ and in each neighborhood, we may choose a reduced representation y of x (Lemma 3.1) and fix a representative $\tilde{a}$ of a so that the zeroes of $\langle x,a\rangle$ in this neighborhood become exactly the zeroes of the single-valued holomorphic function

$\langle y, \tilde{a}\rangle$. However, in the interest of notational simplicity, we prefer to abuse the language a bit and refer to $f_a \equiv \langle x, a\rangle$ itself as a holomorphic function on V even though we actually have in mind such $\langle y, \tilde{a}\rangle$'s. Now it is self-evident that $n_0(s_j, a_j) \leq n_0(t, a_j)$ for all j since $s_j \leq t$. Because $a_j \to a$, f_{a_j} converges to f uniformly on compact sets. So when j is sufficiently large, Hurwitz's theorem implies that

$$\begin{aligned} n_0(t, a_j) &= \sum_{p \in V[t]} (\text{order of zero of } f_{a_j}(p)) \\ &\leq \sum_{p \in V[t]} (\text{order of zero of } f_a(p)) \\ &= n_0(t, a). \end{aligned}$$

Altogether, $n_0(s_j, a_j) \leq n_0(t, a)$ for j sufficiently large. This clearly disposes of (5.4). As for (5.3), we have to prove it by a contradiction. Assume it is false, then $\limsup_{i\to\infty} n_0(r_i, a_i) > n_0(t, a)$. Passing to a subsequence if necessary, we may assume that we have $\lim_{i\to\infty} n_0(r_i, a_i) > n_0(t, a)$, where $r_1 \geq r_2 \geq \cdots \geq t$ and furthermore, that $V[r_1]$ encloses the same number of zeroes of f_a as does $V[t]$, i.e., $n_0(r_1, a) = n_0(t, a)$. Again by Hurwitz's theorem, $n_0(r_1, a_i) \leq n_0(r_1, a)$ whenever i is sufficiently large. Combined with the above, we have $n_0(r_1, a_i) \leq n_0(t, a)$. Since $r_1 \geq r_i$ obviously implies $n_0(r_i, a_i) \leq n_0(r_1, a_i)$, we now have $n_0(r_i, a_i) \leq n_0(t, a)$. Thus we have arrived at a subsequence $\{n_0(r_i, a_i)\}$ such that each member is smaller than or equal to $n_0(t, a)$. This contradicts $\lim_{i\to\infty} n_0(r_i, a_i) > n_0(t, a)$ and proves (5.3). Q.E.D.

§2. We are now ready for the integration. Let Ω be the volume form on $P_n\mathbb{C}$ as in §1. We choose an integrable function β on $P_n\mathbb{C}$ satisfying two conditions:

$$\int_{P_n\mathbb{C}} \beta\Omega = 1 \tag{5.5}$$

$$\beta \geq 0 \tag{5.6}$$

We will specify β later, but for the moment, we simply use these two properties of β and integrate the following with respect to $\beta\Omega$ over $P_n\mathbb{C}$:

$$N_0(r,a) < T_0(r) + c_0,$$

where $a \in P_n\mathbb{C}$ and c_0 is independent of r and a. Hence

$$\int_{P_n\mathbb{C}} N_0(r,a)\beta(a)\Omega(a) < T_0(r) + c_0$$

because of (5.5). Now $N_0(r,a) = \int_{r_0}^{r} n_0(t,a)dt$ and by Lemma 5.3, $n(t,a)$ is nonnegative and measurable on $[r_0,r] \times P_n\mathbb{C}$. Furthermore by (5.5), β is integrable and nonnegative, so Fubini's theorem applies and we have:

$$\int_{r_0}^{r} dt \int_{P_n\mathbb{C}} n_0(t,a)\beta(a)\Omega(a) < T_0(r) + c_0.$$

By Theorem 5.2, this is equivalent to

$$\int_{r_0}^{r} dt \int_{\mathbb{C}^{n+1}} n_0(t,\pi(Z))\beta(\pi(Z))e^{-\langle Z,Z\rangle} dL < CT_0(r) + C' \tag{5.7}$$

where C and C' are positive constants. We want to evaluate the inside integral on the left. Choose an <u>arbitrary</u> O.N. basis $\{e_0,\ldots,e_n\}$ of $\mathbb{C}^{n+1}$, and let $W = \mathrm{span}\{e_0\}$, $W^{\perp} = \mathrm{span}\{e_1,\ldots,e_n\}$ We emphasize the fact that we start off with an arbitrary O.N. basis rather than with the canonical basis $\{\epsilon_0,\ldots,\epsilon_n\}$ of $\mathbb{C}^{n+1}$. Let $\tilde{x}: V \to \mathbb{C}^{n+1}$ be the map which induces $x: V \to P_n\mathbb{C}$ as usual, and let $(y_0,\ldots,y_n)$ be the coordinate functions of $\tilde{x}$ relative to $\{e_0,\ldots,e_n\}$, i.e., for $p \in V$, $\tilde{x}(p) = \Sigma_{A=0}^{n} y_A(p)e_A$. Let $V_0[r]$ be the complement in $V[r]$ of the zeroes of y_0. Then $V[r] - V_0[r]$ is a finite set which includes the common zeroes of $\{y_0,\ldots,y_n\}$. For $Z_1 = (\Sigma_{i=1}^{n} z_i e_i) \in W^{\perp} - \{0\}$, define a holomorphic function $g_{Z_1}: V_0[t] \to \mathbb{C}$ by

$$g_{Z_1}(p) = -\frac{\bar{z}_1 y_1(p)+\cdots+\bar{z}_n y_n(p)}{y_0(p)} = -\frac{\langle\tilde{x}(p),Z_1\rangle}{y_0(p)}$$

Let $\tilde{n}(t,z_0)$ denote the number of pre-images of $\bar{z}_0$ under g_{Z_1}, counting multiplicities. We claim: if we denote by Z the vector $\Sigma_{A=0}^{n} z_A e_A \in \mathbb{C}^{n+1}$, then

$$n_0(t,\pi(Z)) = \tilde{n}(t,z_0).$$

Let $q_1,\ldots,q_\ell$ be the points in $V_0[t]$ such that $\langle\tilde{x}(q_j),Z\rangle = 0$ and let α_j be the unique positive integer such that if ζ_j is a local coordinate function centered at q_j, then $\zeta_j^{-\alpha_j}\langle\tilde{x}(q_j),Z\rangle \neq 0,\infty$. By definition, $\Sigma_j\alpha_j$ is $n_0(r,\pi(Z))$. But clearly, $\langle\tilde{x}(q_j),Z\rangle = 0$ if and only if $\bar{z}_0 y_0(q_j) + \cdots + \bar{z}_n y_n($ $= 0$, if and only if $g_{Z_1}(q_j) = \bar{z}_0$. So $q_1,\ldots,q_\ell$ are exactly the preimages in $V_0[r]$ of $\bar{z}_0$ under g_{Z_1}. Furthermore, the multiplicity with which g_{Z_1} covers $\bar{z}_0$ at q_j is exactly that positive integer β_j such that $\zeta_j^{-\beta_j}(g_{Z_1}(q_j) - \bar{z}_0) \neq 0,\infty$, and this is true if and only if $\zeta_j^{-\beta_j}(-\bar{z}_0 y_0(q_j) - \cdots - \bar{z}_n y_n(q_j))$ $\neq 0,\infty$. So it is clear that $\alpha_j = \beta_j$, and we have proved our claim.

In view of (5.2), we have:

$$n_0(t,\pi(Z))dL = (\frac{\sqrt{-1}}{2}\tilde{n}(t,z_0)dz_0 \wedge d\bar{z}_0)$$
$$\wedge (\frac{\sqrt{-1}}{2})^n(dz_1 \wedge d\bar{z}_1 \wedge \cdots \wedge dz_n \wedge d\bar{z}_n)$$

where $Z = \Sigma_{A=0}^n z_A e_A$ as above. Let us further agree to let dL_1 stand for the volume form

$$(\frac{\sqrt{-1}}{2})^n(dz_1 \wedge d\bar{z}_1) \wedge \cdots \wedge (dz_n \wedge d\bar{z}_n)$$

of $W^\perp$. So the integrand of (5.7) becomes

$$(\frac{\sqrt{-1}}{2}\tilde{n}(t,z_0)\beta(\pi(Z))e^{-|z_0|^2}dz_0 \wedge d\bar{z}_0) \wedge e^{-\langle Z_1,Z_1\rangle}dL_1$$

and by Fubini's theorem:

$$(5.8)\quad \int_{\mathbb{C}^{n+1}} n_0(t,\pi(Z))\beta(\pi(Z))e^{-\langle Z,Z\rangle}dL$$
$$= \int_{W^\perp} e^{-\langle Z_1,Z_1\rangle}dL_1 \cdot \int_{\mathbb{C}} \frac{\sqrt{-1}}{2}\tilde{n}(t,z_0)\beta(\pi(Z))e^{-|z_0|^2}dz_0 \wedge d\bar{z}_0$$

By Lemma 2.12, the last integral over $\mathbb{C}$ is equal to

$$\int_{V_0[t]} g_{Z_1}^*(\frac{\sqrt{-1}}{2}\beta(\pi(z_0 e_0 + Z_1))\exp(-|z_0|^2)dz_0 \wedge d\bar{z}_0)$$

where $g_{Z_1}: V_0[t] \to \mathbb{C}$, and $V_0[t]$ is the complement in $V[t]$

of the zeroes of y_0. Let us evaluate this integrand. So pick a $p \in V_0[t]$ and let a coordinate function ζ be introduced around p. Then in this coordinate neighborhood,

$$g^*_{Z_1}\left(\frac{\sqrt{-1}}{2}\,\beta(\pi(-z_0 e_0 + Z_1))\exp(-|z_0|^2)dz_0 \wedge d\bar{z}_0\right)$$

$$= \frac{\sqrt{-1}}{2}\,\beta\left(\pi\left(-\frac{\langle\tilde{x},Z_1\rangle}{y_0}e_0 + Z_1\right)\right)\cdot\exp\left(-\left|\frac{\langle\tilde{x},Z_1\rangle}{y_0}\right|^2\right)$$

$$\cdot\left|\frac{y_o\Sigma^n_{i=1}\bar{z}_i y'_i - y'_o\Sigma^n_{i=1}\bar{z}_i y_i}{y_0^2}\right|^2 d\zeta\wedge d\bar{\zeta}$$

where we have written y'_A for $\frac{dy_A}{d\zeta}$. If following (3.2), we write $\tilde{x}^{(1)}$ for $\left(\frac{dx_0}{d\zeta},\ldots,\frac{dx_n}{d\zeta}\right)$, the above then becomes:

$$\frac{\sqrt{-1}}{2}\,\beta\left(\pi\left(-\frac{\langle\tilde{x},Z_1\rangle}{y_0}e_0 + Z_1\right)\right)\cdot\exp\left(-\left|\frac{\langle\tilde{x},Z_1\rangle}{y_0}\right|^2\right)$$

$$\cdot\ \frac{|\langle\tilde{x}^{(1)},Z_1\rangle - y_0^{-1}y'_0\langle\tilde{x},Z_1\rangle|^2}{|y_0|^2}\,d\zeta\wedge d\zeta$$

Now substitute this into (5.8) and apply Fubini's theorem once more, we obtain:

$$(5.9)\qquad \int_{\mathbb{C}^{n+1}} n_0(t,\pi(Z))\beta(\pi(Z))e^{-\langle Z,Z\rangle}dL = \int_{V_0[t]}\nu$$

where ν is a C^∞ two form on $V_0[t]$ such that if we write locally $\nu = \frac{\sqrt{-1}}{2}h\,d\zeta\wedge d\bar{\zeta}$, then for each p of $V_0[t]$ in this neighborhood,

$$h(p) = \int_{Z_1\in W^\perp}\beta\left(\pi\left(-\frac{\langle\tilde{x}(p),Z_1\rangle}{y_0}e_0 + Z_1\right)\right)$$

$$\cdot\left|\frac{\langle\tilde{x}^{(1)}(p),Z_1\rangle - y_0^{-1}(p)y'_0(p)\langle\tilde{x}(p),Z_1\rangle}{y_0}\right|^2$$

$$\cdot \exp\{-\langle Z_1,Z_1\rangle - \left|\frac{\langle\tilde{x}(p),Z_1\rangle}{y_0(p)}\right|^2\}dL_1$$

Now recall that we had started off with an arbitrary O.N. basis $\{e_0,\ldots,e_n\}$ of $\mathbb{C}^{n+1}$ and $W^\perp = \mathrm{span}\{e_1,\ldots,e_n\}$. Thus the last integral formula for $h(p)$ is valid for <u>any</u> n-dimensional subspace $W^\perp$ of $\mathbb{C}^{n+1}$. Our task now is to make a judicious choice of $W^\perp$, each choice being dependent on p, so that the integrand assumes its simplest form. Now $p \in V_0[t]$, so by definition $y_0(p) \neq 0$; in particular $\tilde{x}(p) \neq 0$ and the orthogonal complement of $\tilde{x}(p)$ in $\mathbb{C}^{n+1}$ is therefore an n-dimensional subspace $\tilde{x}(p)^\perp$ of $\mathbb{C}^{n+1}$. <u>For this</u> p, <u>we choose</u> $W^\perp$ <u>to be</u> $\tilde{x}(p)^\perp$. With this choice, it follows that $\langle\tilde{x}(p),Z_1\rangle = 0$ for every $Z_1 \in W^\perp$. What is more, $|y_0(p)| = |\tilde{x}(p)|$ if one recalls that $y_0(p)$ is nothing but the component of $\tilde{x}(p)$ in the direction orthogonal to $W^\perp$. Hence the above integral formula for $h(p)$ simplifies to:

$$h(p) = \int_{Z_1\in\tilde{x}(p)^\perp} \beta(\pi(Z_1))\frac{|\tilde{x}^{(1)}(p),Z_1|^2}{|\tilde{x}(p)|^2}\exp(-\langle Z_1,Z_1\rangle)dL_1$$

At this point, we can change notation a bit without fear of confusion. We let dL be the generic symbol of the Lebesgue measure in all complex euclidean spaces and let Z be the generic symbol of the coordinate function in the same spaces. We can then rewrite the above in a more civilized fashion:

$$(5.10)\quad h(p) = \int_{Z\in\tilde{x}(p)^\perp} \beta(\pi(Z))\frac{|\tilde{x}^{(1)}(p),Z|^2}{|\tilde{x}(p)|^2}\exp(-\langle Z,Z\rangle)dL$$

or more simply:

$$(5.11)\quad h = \int_{\tilde{x}^{\perp}} \beta(\pi(Z)) \frac{|\tilde{x}^{(1)},Z|^2}{|\tilde{x}|^2} \exp(-\langle Z,Z\rangle)dL$$

Return now to (5.9). Using the notation there, we have by virtue of (5.7) that

$$\int_{r_o}^{r} dt \int_{V_o[t]} \nu < CT_o(r) + C',$$

where C and C' are positive constants. Since $V[t]$ and $V_o[t]$ differ by a finite point set, we may replace $\int_{V_o[t]} \nu$ by $\int_{V[t]} \nu$. Furthermore (5.6) and (5.11) imply that $h \geq 0$, and since locally $\nu = \frac{\sqrt{-1}}{2} h\, d\zeta \wedge d\bar{\zeta}$, we see that

$$\int_{V[t]-V[r_o]} \nu \leq \int_{V[t]} \nu,$$

so that

$$\int_{r_o}^{r} dt \int_{V[t]-V[r_o]} \nu < CT_o(r) + C'.$$

But in $V - V[r_o]$, τ is harmonic and we can locally introduce the holomorphic function $\sigma = \tau + \sqrt{-1}\,\rho$ as a coordinate function, except on the discrete set of critical points of τ. Therefore, except on a finite point set of $V[t] - V[r_o]$, we can write

$$\nu = \frac{\sqrt{-1}}{2} h\, d\sigma \wedge d\bar{\sigma} = h\, d\tau \wedge * d\tau$$

with h given by (5.11), with the understanding that $\tilde{x}^{(1)}$

is now given by $(\frac{dx_o}{d\sigma},\ldots,\frac{dx_n}{d\sigma})$. As usual, we ignore such exceptional finite point sets in integration, so that

$$\int_{V[t]-V[r_o]} \nu = \int_{V[t]-V[r_o]} h\, d\tau \wedge *d\tau$$
$$= \int_{r_o}^{t} ds \int_{\partial V[s]} h * d\tau ,$$

where the last step is due to Fubini's theorem and the existence of special coordinate functions (Definition 2.2). Putting these together, we have the following:

$$(5.12) \qquad \int_{r_o}^{r} dt \int_{r_o}^{t} ds \int_{\partial V[s]} h * d\tau < CT_o(r) + C'$$

where C, C' are positive constants and if we write

$$\tilde{x}^{(1)} = (\frac{dx_o}{d\sigma},\ldots,\frac{dx_n}{d\sigma}),$$

then

$$h = \int_{\tilde{x}^{\perp}} \beta(\pi(Z)) \frac{|\tilde{x}^{(1)},Z|^2}{|\tilde{x}|^2} \exp(-\langle Z,Z\rangle)dL.$$

Thus far, we have not specified β but rather, we have proceeded solely on the assumption that (5.5) and (5.6) are valid for β. We now make an initial step toward the explicit determination of a suitable β by introducing an integrable function φ: $[0,1] \to \mathbb{R}$ and fixing a unit vector $b \in \mathbb{C}^{n+1}$. Then we require that

$$(5.13) \qquad \beta(a) = \varphi(|a,b|^2)$$

for all $a \in P_n\mathbb{C}$. The right side inner product $|a,b|$ can be taken with any unit vector representing a, and is clearly independent of the particular choice of the representative. Then for $Z \in \mathbb{C}^{n+1}$,

$$(5.13)' \qquad \beta(\pi(Z)) = \beta(\frac{Z}{|Z|}) = \varphi(\frac{|Z,b|^2}{|Z|^2}).$$

This choice of β, that it be dependent only on a single coordinate, is motivated by the choice of the density function ρ made in Chapter II, §6 in the case $n = 1$. Now β so defined should still obey (5.5) and (5.6) if (5.12) is to remain valid. We claim that if φ satisfies (5.14) and (5.15) below, then β would indeed satisfy (5.5) and (5.6).

$$(5.14) \qquad \varphi \geq 0$$

$$(5.15) \qquad \int_0^1 \varphi(t)(1 - t)^{n-1}dt = \frac{(n-1)!}{\pi^n}$$

That (5.14) implies (5.6) is obvious. We will show that (5.15) is equivalent to (5.5) by an explicit computation. By Theorem 5.2, (5.5) is equivalent to:

$$\int_{\mathbb{C}^{n+1}} \beta(\pi(Z))e^{-\langle Z,Z\rangle}dL = n!\pi$$

By (5.13)', this equals

$$\int_{\mathbb{C}^{n+1}} \varphi(\frac{|Z,b|^2}{|Z|^2})e^{-\langle Z,Z\rangle}dL = n!\pi.$$

Now the Lebesgue integral on $\mathbb{C}^{n+1}$ is invariant under the unitary group. So if we move the unit vector b to $\epsilon_0 = (1,0,\ldots,0)$ by a unitary transformation, the above may be rewritten as:

$$\int_{\mathbb{C}^{n+1}} \varphi\left(\frac{z_0\bar{z}_0}{z_0\bar{z}_0+\cdots+z_n\bar{z}_n}\right)e^{-(z_0\bar{z}_0+\cdots+z_n\bar{z}_n)}\left(\frac{\sqrt{-1}}{2}\,dz_0\wedge d\bar{z}_0\right)\wedge\cdots$$

$$\wedge\left(\frac{\sqrt{-1}}{2}\,dz_n\wedge dz_n\right) = n!\pi.$$

Introduce polar coordinates: $z_A = r_A e^{\sqrt{-1}\theta_A}$, then

$$\frac{\sqrt{-1}}{2}\,dz_A\wedge d\bar{z}_A = r_A\,dr_A\wedge d\theta_A\,.$$

Let $t_A = r_A^2$, then the above becomes:

$$\int_0^\infty\cdots\int_0^\infty \varphi\left(\frac{t_0}{t_0+\cdots+t_n}\right)e^{-t_0-\cdots-t_n}dt_0\cdots dt_n = \frac{n!}{\pi^n}\,.$$

Now the mapping $(\tau, s_1, \ldots, s_n) \to (t_0, \ldots, t_n)$ such that

$$t_0 = \tau(s_1 + \cdots + s_n)$$

$$t_1 = (1-\tau)s_1$$

$$\cdots\cdots\cdots$$

$$t_n = (1-\tau)s_n$$

clearly maps $(0,1)\times(0,\infty)\times\cdots\times(0,\infty)$ one-one and onto $(0,\infty)\times\cdots\times(0,\infty)$ ((n+1)-times). By a simple induction argument:

$$dt_0\wedge\cdots\wedge dt_n = (1-\tau)^{n-1}(s_1+\cdots+s_n)d\tau\wedge ds_1\wedge\cdots\wedge ds_n.$$

Observe further that $(t_0+\cdots+t_n) = (s_1+\cdots+s_n)$, so that $\tau = \frac{t_0}{t_0+\cdots+t_n}$. Hence we may transform the above integral into:

$$\int_0^1 \varphi(\tau)(1-\tau)^{n-1}d\tau \cdot \int_0^\infty \cdots \int_0^\infty (s_1+\cdots+s_n)e^{-s_1-\cdots-s_n}ds_1\cdots ds_n = \frac{n!}{\pi^n},$$

$$\Longleftrightarrow \int_0^1 \varphi(\tau)(1-\tau)^{n-1}d\tau \cdot (\Sigma_{i=1}^n \int_0^\infty \cdots \int_0^\infty s_i e^{-s_1-\cdots-s_n}ds_1\cdots ds_n) = \frac{n!}{\pi^n},$$

$$\Longleftrightarrow \int_0^1 \varphi(\tau)(1-\tau)^{n-1}d\tau \{n(\int_0^\infty e^{-s}ds)^{n-1} \cdot \int_0^\infty se^{-s}ds\} = \frac{n!}{\pi^n}.$$

Because $\int_0^\infty e^{-s}ds = 1$ and $\int_0^\infty se^{-s}ds = \Gamma(2) = 1$, this is equivalent to $n \cdot \int_0^1 \varphi(\tau)(1-\tau)^{n-1}ds = \frac{n!}{\pi^n}$, which is precisely (5.15).

We now return to the computation of h in (5.12). Because of (5.13)', we can write: for any $p \in V[t] - V[r_0]$ which is not among the finite set of critical points of τ or the zeroes of $\tilde{x}$,

$$h(p) = \int_{\tilde{x}(p)^\perp} \exp(-\langle Z,Z\rangle) \cdot \varphi(\frac{|b,Z|^2}{|Z|^2}) \cdot \frac{|\tilde{x}^{(1)}(p),Z|^2}{|\tilde{x}(p)|^2}\, dL$$

where it is understood that φ satisfies (5.14) and (5.15). Let us choose O.N. basis $e_0,\ldots,e_n$ in $\mathbb{C}^{n+1}$ so that $\text{span}\{e_0\} = \text{span}\{\tilde{x}(p)\}$, $\text{span}\{e_0,e_1\} = \text{span}\{\tilde{x}(p),b\}$, and $\text{span}\{e_0,e_1,e_2\} = \text{span}\{\tilde{x}(p),b,\tilde{x}^{(1)}(p)\}$. Then $\tilde{x}(p)^\perp = \text{span}\{e_1,\ldots,e_n\}$. For simplicity, we will write $\Sigma_{A=0}^n z_A e_A$ as $(z_0,\ldots,z_n)$. Thus in this notation, we may write:

$$\tilde{x}(p) = (a_1,0,\ldots,0)$$
$$b = (a_2,a_3,0,\ldots,0)$$
$$\tilde{x}^{(1)}(p) = (a_4,a_5,a_6,0,\ldots,0).$$

With this simplification, and noting that $Z \in \tilde{x}(p)^{\perp}$ implies $Z \doteq (0, z_1, \ldots, z_n)$, we have:

$$h(p) = \int_{\tilde{x}(p)^{\perp}} \exp(-z_1\bar{z}_1 - \cdots - z_n\bar{z}_n) \cdot \varphi\left(\frac{z_1\bar{z}_1 \cdot a_3\bar{a}_3}{z_1\bar{z}_1 + \cdots + z_n\bar{z}_n}\right) \cdot \frac{|a_5 z_1 + a_6 z_2|^2}{|a_1|^2}$$

We have $|a_5 z_1 + a_6 z_2|^2 = |a_5 z_1|^2 + |a_6 z_2|^2 + a_5\bar{a}_6 z_1\bar{z}_2 + a_6\bar{a}_5 z_2\bar{z}_1$. The integral over $\tilde{x}(p)^{\perp}$ of the last two terms must vanish because of antisymmetry considerations. Hence

$$\begin{aligned} h(p) &= \int_{\tilde{x}(p)^{\perp}} \exp(-z_1\bar{z}_1 - \cdots - z_n\bar{z}_n) \cdot \varphi\left(\frac{|z_1|^2 |a_3|^2}{z_1\bar{z}_1 + \cdots + z_n\bar{z}_n}\right) \\ &\qquad \cdot \frac{|a_5 z_1|^2 + |a_6 z_2|^2}{|a_1|^2}\, dL \\ &\geq \int_{\tilde{x}(p)^{\perp}} \exp(-z_1\bar{z}_1 - \cdots - z_n\bar{z}_n) \cdot \varphi\left(\frac{|z_1|^2 |a_3|^2}{z_1\bar{z}_1 + \cdots + z_n\bar{z}_n}\right) \\ &\qquad \cdot \frac{|a_5 z_1|^2}{|a_1|^2}\, dL \\ &= \frac{|a_5|^2}{|a_1|^2} \int_{\tilde{x}(p)^{\perp}} \exp(-z_1\bar{z}_1 - \cdots - z_n\bar{z}_n) \cdot \varphi\left(\frac{|z_1|^2 |a_3|^2}{z_1\bar{z}_1 + \cdots + z_n\bar{z}_n}\right) \\ &\qquad \cdot |z_1|^2 dL \end{aligned}$$

Writing $f(a_3)$ for the last integral, we have:

$$\frac{|a_5|^2}{|a_1|^2} f(a_3) \leq h(p) \tag{5.16}$$

We claim:

$$f(a_3) = \begin{cases} \pi\varphi(|a_3|^2) & \text{if } n = 1 \\ n(n-1)\pi^n \int_0^1 t\varphi(|a_3|^2 t)(1-t)^{n-2} dt & \text{if } n \geq 2. \end{cases} \tag{5.17}$$

Let us prove the claim for $n \geq 2$. To this end, employ polar coordinates $z_i = r_i e^{\sqrt{-1}\theta}$ as before, and let $t_i = r_i^2$. Then

$$f(a_3) = \pi^n \int_0^\infty \cdots \int_0^\infty t_1 e^{-t_1 - \cdots - t_n} \varphi\left(\frac{t_1 |a_3|^2}{t_1 + \cdots + t_n}\right) dt_1 \cdots dt_n$$

Then use the same transformation $(\tau, s_2, \ldots, s_n) \to (t_1, \ldots, t_n)$ such that $t_1 = \tau(s_2 + \cdots + s_n)$, $t_2 = (1-\tau)s_2, \ldots, t_n = (1-\tau)s_n$, to map $(0,1) \times (0,\infty) \times \cdots \times (0,\infty)$ diffeomorphically onto $(0,\infty) \times \cdots \times (0,\infty)$ (n times). Observing as before that

$$dt_1 \wedge \cdots \wedge dt_n = (1-\tau)^{n-2}(s_2 + \cdots + s_n) d\tau \wedge ds_2 \wedge \cdots \wedge ds_n,$$

we see that

$$\begin{aligned} f(a_3) &= \pi^n \int_0^1 \tau(1-\tau)^{n-2} \varphi(\tau |a_3|^2) d\tau \\ &\quad \cdot \int_0^\infty \cdots \int_0^\infty e^{-s_2 - \cdots - s_n} (s_2 + \cdots + s_n)^2 ds_2 \cdots ds_n \\ &= \pi^n \int_0^1 \tau(1-\tau)^{n-2} \varphi(\tau |a_3|^2) d\tau \\ &\quad \cdot \Big\{ \int_0^\infty \cdots \int_0^\infty e^{-s_2 - \cdots - s_n} (s_2^2 + \cdots + s_n^2) ds_2 \cdots ds_n \\ &\quad + 2 \int_0^\infty \cdots \int_0^\infty \Sigma_{i<j}\, s_i s_j e^{-s_2 - \cdots - s_n} ds_2 \cdots ds_n \Big\} \\ &= \pi^n \int_0^1 \tau(1-\tau)^{n-2} \varphi(\tau |a_3|^2) d\tau \\ &\quad \cdot \Big\{ (n-1) \int_0^\infty s^2 e^{-s} ds \cdot \left(\int_0^\infty e^{-s} ds \right)^{n-2} \\ &\quad + 2 \frac{(n-1)(n-2)}{2} \left(\int_0^\infty s e^{-s} ds \right)^2 \left(\int_0^\infty e^{-s} ds \right)^{n-4} \Big\}. \end{aligned}$$

But $\int_0^\infty e^{-s} ds = \int_0^\infty s e^{-s} ds = 1$, and $\int_0^\infty s^2 e^{-s} ds = \Gamma(3) = 2$, so we get:

$$f(a_3) = \pi^r \int_0^1 \tau(1-\tau)^{n-2}\varphi(\tau|a_3|^2)d\tau\{2(n-1)+(n-1)(n-2)\}$$
$$= n(n-1)\pi^n \int_0^1 \tau(1-\tau)^{n-2}\varphi(\tau|a_3|^2)d\tau,$$

which is exactly (5.17).

(5.17) prompts us to introduce a function $\psi: [0,1] \to \mathbb{R}$ such that

$$(5.18) \qquad \psi(s) = \begin{cases} \pi\varphi(s) & \text{if} \quad n = 1 \\ \frac{\pi^n}{(n-2)!}\int_0^1 t\varphi(st)(1-t)^{n-2}dt & \text{if} \quad n \geq 2 . \end{cases}$$

($\frac{\pi^n}{(n-2)!}$ is a normalizing factor to insure (5.22) below). Then combining (5.16)-(5.18) and (5.12), we have:

$$(5.19) \qquad \int_{r_0}^r dt \int_{r_0}^t ds \int_{\partial V[s]} h *d\tau < CT_0(r) + C'$$

where C, C' are positive constants independent of b, and at p, $h(p) = \frac{|a_5|^2}{|a_1|^2}\psi(|a_3|^2)$, provided O.N. basis $e_0,\ldots,e_n$ is so chosen that

$$\tilde{x}(p) = (a_1,0,\ldots,0)$$
$$b = (a_2,a_3,0,\ldots,0)$$
$$\tilde{x}^{(1)}(p) = (a_4,a_5,a_6,0,\ldots,0).$$

We now rewrite this expression of h in intrinsic notation. Relative to the same O.N. basis,

$$\frac{|\langle b \wedge \tilde{x}(p), \tilde{x}(p) \wedge \tilde{x}^{(1)}(p)\rangle|^2}{|\tilde{x}(p)|^4 |b \wedge \tilde{x}(p)|^2} = \frac{|a_1 a_3 \bar{a}_1 \bar{a}_5|^2}{|a_1|^4 |a_1 a_3|^2} = \frac{|a_5|^2}{|a_1|^2},$$

$$\frac{|b \wedge \tilde{x}(p)|^2}{|\tilde{x}(p)|^2} = \frac{|a_1 a_3|^2}{|a_1|^2} = |a_3|^2 \quad .$$

So, we may rewrite (5.19) as:

$$(5.20) \quad \int_{r_0}^{r} dt \int_{r_0}^{s} ds \int_{\partial V[s]} \frac{|\langle b \wedge \tilde{x}, \tilde{x} \wedge \tilde{x}^{(1)} \rangle|^2}{|\tilde{x}|^4 |b \wedge \tilde{x}|^2} \cdot \psi\left(\frac{|b \wedge \tilde{x}|^2}{|\tilde{x}|^2}\right) * d\tau$$
$$< CT_0(r) + C'$$

where C, C' are positive constants independent of b, ψ is given by (5.18) and $\tilde{x}^{(1)} = (\frac{dx_0}{d\sigma}, \ldots, \frac{dx_n}{d\sigma})$.

Recall that we began the integration with an integrable function β on $P_n\mathbb{C}$ satisfying (5.5) and (5.6) and, then we determined β in terms of a real-valued function $\varphi\colon [0,1] \to \mathbb{R}$ (cf. (5.13)). We proved that so long as φ satisfies (5.14) and (5.15), β will satisfy (5.5) and (5.6). Otherwise φ is arbitrary. In (5.18), we defined a new function $\psi\colon [0,1] \to \mathbb{R}$ in terms of φ. Now we make this claim: ψ determines φ uniquely, and if ψ satisfies the following two conditions, then φ would satisfy (5.14) and (5.15):

(5.21) The derivatives of ψ exist on $(0,1)$ and are positive there.

$$(5.22) \qquad \int_0^1 \psi(s)ds = 1.$$

Granting this claim for the moment, we bring our search for a suitable β satisfying (5.5) and (5.6) to an end by locating a suitable ψ such that (5.21) and (5.22) are fulfilled. Such a ψ turns out to be exceedingly simple:

$$\psi(s) = (1-\alpha)(1-s)^{-\alpha}, \quad 0 < \alpha < 1.$$

It is easy to see that this ψ satisfies both (5.21) and (5.22). The constant α is arbitrary for the moment and will be allowed to vary later on. On the basis of this choice of ψ, we may rewrite (5.20) as

$$(5.23)\quad (1-\alpha)\int_{r_0}^{r} dt \int_{r_0}^{t} ds \int_{\partial V[s]} \frac{|\langle b\wedge\tilde{x},\tilde{x}\wedge\tilde{x}^{(1)}\rangle|^2}{|\tilde{x}|^4\,|b\wedge\tilde{x}|^2}\left(1-\frac{|b\wedge\tilde{x}|^2}{|\tilde{x}|^2}\right)^{-\alpha} * d\tau$$
$$< CT_0(r) + C'$$

where C, C' are positive constants independent of α and b. Now note that since b is a unit vector, $|Z|^2 - |b\wedge Z|^2 = |b,Z|^2$ for every $Z \in \mathbb{C}^{n+1}$; this is most easily proven by choosing O.N. basis $e_0,\ldots,e_n$ so that $b = e_0$ and $\operatorname{span}\{e_0,e_1\} = \operatorname{span}\{b,Z\}$. This being the case,

$$\left(1-\frac{|b\wedge\tilde{x}|^2}{|\tilde{x}|^2}\right)^{-\alpha} = \left(\frac{|\tilde{x}|^2}{|\tilde{x}|^2-|b\wedge\tilde{x}|^2}\right)^{\alpha} = \left(\frac{|\tilde{x}|^2}{|b,\tilde{x}|^2}\right)^{\alpha} .$$

Furthermore, we have the identity:

$$|\langle b\wedge\tilde{x},\tilde{x}\wedge\tilde{x}^{(1)}\rangle|^2 = |\tilde{x}|^2|b\,\lrcorner\,(\tilde{x}\wedge\tilde{x}^{(1)})|^2 - |b,\tilde{x}|^2|\tilde{x}\wedge\tilde{x}^{(1)}|^2$$

where $\lrcorner$ denotes the interior product of Chapter I §3. Again this identity is easily proven pointwise at each p of $V[t] - V[r_0]$ by choosing O.N. basis as in (5.19). It should be pointed out that the above does not make sense at those (finite number of) p's where $\tilde{x}$ vanishes, but since one need only worry about the behavior of the integrand almost everywhere, such exception may be ignored. Using these two identities, the integrand of (5.23) is:

$$\frac{|\langle b\wedge\tilde{x},\tilde{x}\wedge\tilde{x}^{(1)}\rangle|^2}{|\tilde{x}|^4|b\wedge\tilde{x}|^2}\left(1-\frac{|b\wedge\tilde{x}|^2}{|\tilde{x}|^2}\right)^{-\alpha}$$

$$=\frac{|\tilde{x}|^2|b\lrcorner(\tilde{x}\wedge\tilde{x}^{(1)})|^2-|b,\tilde{x}|^2|\tilde{x}\wedge\tilde{x}^{(1)}|^2}{|\tilde{x}|^4|b\wedge\tilde{x}|^2}\cdot\left(\frac{|\tilde{x}|}{|b,\tilde{x}|}\right)^{2\alpha}$$

$$\geq\frac{|\tilde{x}|^2|b\lrcorner(\tilde{x}\wedge\tilde{x}^{(1)})|^2-|b,\tilde{x}|^2|\tilde{x}\wedge\tilde{x}^{(1)}|^2}{|\tilde{x}|^4|\tilde{x}|^2}\left(\frac{|\tilde{x}|}{|b,\tilde{x}|}\right)^{2\alpha}$$

(because $|b\wedge\tilde{x}|^2\leq|\tilde{x}|^2|b|^2=|\tilde{x}|^2$ by (1.11))

$$>\frac{|b\lrcorner(\tilde{x}\wedge\tilde{x}^{(1)})|^2}{|\tilde{x}|^4}\cdot\left(\frac{|\tilde{x}|}{|b,\tilde{x}|}\right)^{2\alpha}-\frac{|\tilde{x}\wedge\tilde{x}^{(1)}|^2}{|\tilde{x}|^4},$$

and the last step is because $\left(\frac{|\tilde{x}|}{|b,\tilde{x}|}\right)^{2\alpha}\leq\left(\frac{|\tilde{x}|}{|b,\tilde{x}|}\right)^2$ since $\alpha<1$ and Schwarz's inequality (1.10) says $\frac{|\tilde{x}|}{|b,\tilde{x}|}\geq 1$. (5.23) therefore simplifies to:

$$(1-\alpha)\int_{r_o}^{r}dt\int_{r_o}^{t}ds\int_{\partial V[s]}\frac{|b\lrcorner(\tilde{x}\wedge\tilde{x}^{(1)})|^2}{|\tilde{x}|^4}\left(\frac{|\tilde{x}|}{|b,\tilde{x}|}\right)^{2\alpha}*d\tau$$

$$\leq(1-\alpha)\int_{r_o}^{r}dt\int_{r_o}^{t}ds\int_{\partial V[s]}\frac{|\tilde{x}\wedge\tilde{x}^{(1)}|^2}{|\tilde{x}|^4}*d\tau+CT_o(r)+C'$$

$$=(1-\alpha)\int_{r_o}^{r}dt\int_{V[t]-V[r_o]}\frac{|\tilde{x}\wedge\tilde{x}^{(1)}|^2}{|\tilde{x}|^4}d\tau\wedge*d\tau+CT_o(r)+C'$$

where it may be recalled $\tilde{x}^{(1)}=\left(\frac{dx_o}{d\sigma},\ldots,\frac{dx_n}{d\sigma}\right)$. Now according to Remark (2) following Theorem 4.8, we have:

$$\frac{|\tilde{x}\wedge\tilde{x}^{(1)}|^2}{|\tilde{x}|^4}d\tau\wedge*d\tau=\frac{\sqrt{-1}}{2}\frac{|\tilde{x}\wedge\tilde{x}^{(1)}|^2}{|\tilde{x}|^4}d\sigma\wedge d\bar{\sigma}=x^*\omega$$

except on a discrete set (which is the critical points of τ and the zeroes of $\tilde{x}$). As usual, such discrete sets are ignored in the integration so that:

$$\begin{aligned}(1-\alpha)\int_{r_0}^{r} dt \int_{V[t]-V[r_0]} \frac{|\tilde{x}\wedge\tilde{x}^{(1)}|^2}{|\tilde{x}|^4}\, d\tau\wedge * d\tau &= (1-\alpha)\int_{r_0}^{r} dt \int_{V[t]-V[r_0]} x^*\omega \\ &< (1-\alpha)\int_{r_0}^{r} dt \int_{V[t]} x^*\omega \\ &= (1-\alpha)T_0(r) \\ &< T_0(r).\end{aligned}$$

Using this, we may rewrite the previous inequality in its final form, which is inequality (32) in Ahlfors' paper [1]:

$$(5.24)\quad (1-\alpha)\int_{r_0}^{r} dt \int_{r_0}^{t} ds \int_{\partial V[s]} \frac{|b \lrcorner (\tilde{x}\wedge\tilde{x}^{(1)})|^2}{|\tilde{x}|^4} \left(\frac{|\tilde{x}|}{|b,\tilde{x}|}\right)^{2\alpha} * d\tau < CT_0(r) + C',$$

where $0 < \alpha < 1$, C, C' are positive constants independent of b and α and in terms of the special coordinate function $\sigma = \tau + \sqrt{-1}\rho$, $\tilde{x}^{(1)} = (\frac{dx_0}{d\sigma},\ldots,\frac{dx_n}{d\sigma})$. _We note that_ C _and_ C' _will be our generic symbols of positive constants._ Although in the sequel, we will denote by the same letters C and C' the two constants that appear in the right side of the many variants of (5.24), these constants actually change their values several times.

To round off the proof of (5.24), we must now show that ψ determines φ uniquely and that if ψ satisfies (5.21) and (5.22), then indeed φ satisfies (5.14) and (5.15). This is not difficult. For by (5.18),

$$\psi(s) = \frac{\pi^n}{(n-2)!}\int_0^1 t\varphi(st)(1-t)^{n-2}dt .$$

Letting $\tau = st$, we get,

$$\psi(s) = \frac{\pi^n}{(n-2)!}\int_0^s \frac{1}{s^n}\tau\varphi(\tau)(s-\tau)^{n-2}d\tau ,$$

or equivalently,

$$s^n\psi(s) = \frac{\pi^n}{(n-2)!}\int_0^s \tau(s-\tau)^{n-2}\varphi(\tau)d\tau . \tag{5.25}$$

Recall this formula from calculus:

$$\frac{d}{dx}\int_0^x f(x,y)dy = \int_0^x \frac{df}{dx}(x,y)dy + f(x,y) .$$

We therefore have that

$$\frac{d}{ds}(s^n\psi(s)) = \frac{\pi^n}{(n-3)!}\int_0^s \tau\varphi(\tau)(s-\tau)^{n-3}d\tau ,$$

$$\cdots\cdots\cdots\cdots\cdots\cdots$$

$$\frac{d^{n-2}}{ds^{n-2}}(s^n\psi(s)) = \pi^n s\varphi(s),$$

or,

$$\varphi(s) = \frac{1}{\pi^n s}\frac{d^{n-2}}{ds^{n-2}}(s^n\psi(s)) .$$

Thus ψ completely determines φ and this formula also shows that (5.21) implies (5.14). We now show that (5.22) is equivalent to (5.15). By (5.25) and Fubini's theorem:

$$\begin{aligned}\int_0^1 \psi(s)ds &= \int_0^1 ds\{\frac{\pi^n}{(n-2)!}\frac{1}{s^n}\int_0^s \tau(s-\tau)^{n-2}\varphi(\tau)d\tau\} \\ &= \frac{\pi^n}{(n-2)!}\int_0^1 \tau\varphi(\tau)d\tau\int_\tau^1 (s-\tau)^{n-2}\frac{ds}{s^n} \\ &= \frac{\pi^n}{(n-2)!}\int_0^1 \tau\varphi(\tau)d\tau\int_\tau^1 (1-\frac{\tau}{s})^{n-2}\frac{ds}{s^2}\end{aligned}$$

$$= \frac{\pi^n}{(n-2)!}\int_0^1 \tau\varphi(\tau)d\tau \cdot \frac{1}{(n-1)\tau}\int_\tau^1 d((1-\frac{\tau}{s})^{n-1})$$

$$= \frac{\pi^n}{(n-1)!}\int_0^1 \varphi(\tau)(1-\tau)^{n-1}d\tau.$$

Our assertion that (5.22) is equivalent to (5.15) is now obvious. Q.E.D.

§3. The goal of this section is to generalize (5.24) to the associated holomorphic curve of rank k of x, and we do this by employing a trick due to Ahlfors. Fix a unit decomposable k-vector A^{k-1}, $k = 0,\ldots,n-1$. Of course, if $k > 0$, A^{k-1} will also stand for a $(k-1)$-dimensional projective subspace of $P_n\mathbb{C}$ by our notational convention. Then the contracted curve of the first kind $X^k \lrcorner A^{k-1}: V \to A^\perp$ ($A^\perp$ = polar space of A^{k-1}) may be defined as in (3.6). We now apply (5.24) to the holomorphic curve $X^k \lrcorner A^{k-1}$, then $\tilde{x}$ should be replaced by $(X^k_\sigma \lrcorner A^{k-1})$, and Lemma 3.9 implies that $(\tilde{x}\wedge\tilde{x}^{(1)})$ should be replaced by $\langle A^{k-1},X^{k-1}_\sigma\rangle(X^{k+1}_\sigma \lrcorner A^{k-1})$. The order function of $X^k \lrcorner A^{k-1}$, by (4.35), is simply $T_k(r,A^{k-1})$. But so far as the <u>inequality</u> (5.24) is concerned, Lemma 4.11 says that we may replace $T_k(r,A^{k-1})$ by $T_k(r) + b_k$, where b_k is a constant independent of A^{k-1}. Thus putting all these together, and <u>agreeing to write</u> X^k <u>for</u> X^k_σ <u>from now on</u>, we obtain:

$$(5.26)\quad (1-\alpha)\int_{r_0}^r dt\int_{r_0}^t ds\int_{\partial V[s]} \frac{|A^{k-1},X^{k-1}|^2|b \lrcorner (X^{k+1}\lrcorner A^{k-1})|^2}{|A^{k-1}\lrcorner X^k|^4} \cdot \left(\frac{|A^{k-1}\lrcorner X^k|}{|b,(A^{k-1}\lrcorner X^k)|}\right)^{2\alpha} * d\tau$$

$$< CT_k(r) + C'$$

where $0 < \alpha < 1$, and C, C' are positive constants independent of α, b and A^{k-1}. This is still not sufficiently general for our purpose. We wish to extend the left side from A^{k-1} to all A^{ℓ}, $-1 \leq \ell \leq k-1$. For this, we must discuss a general averaging process. Let $\Phi(A^h)$ be a function of h-dimensional projective subspaces of $P_n\mathbb{C}$ (i.e. $\Phi : G(n,h) \to \mathbb{R}$) and let us fix an A^{h-1}. We want to take the average of Φ over all $A^h \supset A^{h-1}$. This is done in the following manner. Let $A^{\perp}$ be the polar space of A^{h-1} (so $\dim A^{\perp} = n-h$). Then a moment's reflection will show that for each $A^h \supset A^{h-1}$, there is a unique $a \in A^{\perp}$ such that $A^h = A^{h-1} \wedge a$. Therefore, when we restrict the domain of definition of Φ to such A^h's containing the fixed A^{h-1}, Φ becomes a function of $a \in A^{\perp}$, thus giving rise to a function: $A^{\perp} \to \mathbb{R}$, which we simply denote by $\Phi(A^{h-1} \wedge a)$. The <u>average</u> $\underset{A^h \supset A^{h-1}}{\mathfrak{M}} \Phi(A^h)$ <u>of</u> Φ <u>over all h-dimensional projective spaces containing the fixed</u> A^{h-1} is by definition the arithmetic mean of $\Phi(A^{h-1} \wedge a)$ over $A^{\perp}$. i.e.,

$$\underset{A^h \supset A^{h-1}}{\mathfrak{M}} \Phi(A^h) = \Big(\int_{A^{\perp}} \Omega\Big)^{-1}\Big(\int_{A^{\perp}} \Phi(A^{h-1} \wedge a)\Omega(a)\Big)$$

where Ω denotes the volume form of the F-S metric on the projective space $A^{\perp}$. (The notation agrees with that of (4.26)). In practice we have to lift the domain of integration from $A^{\perp}$ to E, where E is the $(n-h+1)$-dimensional vector subspace of $\mathbb{C}^{n+1}$ corresponding to $A^{\perp}$. Thus by Theorem 5.2, we may equivalently define:

$$(5.27)\qquad \underset{A^h \supset A^{h-1}}{\mathfrak{M}} \Phi(A^h) = \Big(\int_E e^{-\langle Z,Z\rangle} dL\Big)^{-1}\Big(\int_E e^{-\langle Z,Z\rangle}\Phi\big([\tfrac{Z}{|Z|}]\wedge A^{h-1}\big)dL\Big)$$

where dL denotes the Lebesgue measure of E and Z the canonical coordinate function of E. $[\frac{Z}{|Z|}]$ is the point of $A^{\perp}$ represented by $\frac{Z}{|Z|}$. (See Chapter I, §1).

The following lemma is basic.

Lemma 5.4. Let Λ^{ℓ} be a decomposable $(\ell+1)$-vector and let h be such that $\ell \geq h \geq 0$. Then for a fixed A^{h-1},

$$\underset{A^h \supset A^{h-1}}{\mathfrak{M}} \log|A^h \lrcorner \Lambda^{\ell}| = \log|A^{h-1} \lrcorner \Lambda^{\ell}| + a_{h\ell},$$

where $a_{h\ell}$ is a constant independent of A^{h-1} and Λ^{ℓ} and depending only on h and ℓ.

*Proof. Write $\mathbb{C}^{n+1} = \Lambda^{\ell} \oplus \Lambda^{\perp}$ (orthogonal decomposition). This leads to an orthogonal projection $p\colon \mathbb{C}^{n+1} \to \Lambda^{\ell}$. Let F be the h-dimensional vector subspace of $\mathbb{C}^{n+1}$ corresponding to A^{h-1}. If $\dim p(F) < h$, then F contains an element of $\Lambda^{\perp}$. Since $F = A^{h-1}$ as multivectors, in this case, both sides of the above identity clearly equal $-\infty$ and there is nothing to prove. So let $\dim p(F) = h$. We may then choose O.N. basis $e_0,\ldots,e_n$ of $\mathbb{C}^{n+1}$ so that $\{e_0,\ldots,e_{h-1}\}$ is a basis of $p(F)$ and so that $\{e_0,\ldots,e_{\ell}\}$ is a basis of Λ^{ℓ}. Obviously, $A^{h-1} = \alpha e_0 \wedge \cdots \wedge e_{h-1}$ + (terms involving $e_{\ell+1},\ldots,e_n$) and $\Lambda^{\ell} = e_0 \wedge \cdots \wedge e_{\ell}$ so that $A^{h-1} \lrcorner \Lambda^{\ell}$ $= \alpha e_h \wedge \cdots \wedge e_{\ell}$. Hence $|A^{h-1} \lrcorner \Lambda^{\ell}| = |\alpha|$. On the other hand, let a be a unit vector orthogonal to F so that $A^{h-1} \wedge a$ $= A^h$ and let $a = a_0 e_0 + \cdots + a_n e_n$. Then since $A^h \lrcorner \Lambda^{\ell}$

$= (A^{h-1}\wedge a) \lrcorner \Lambda^{\ell} = a \lrcorner (A^{h-1} \lrcorner \Lambda^{\ell}) = a \lrcorner (\alpha e_h \wedge \cdots \wedge e_{\ell})$, we have

$$|A^h \lrcorner \Lambda^{\ell}| = |\alpha|(|a_h|^2 + \cdots + |a_{\ell}|^2)^{1/2}$$
$$= |A^{h-1}\wedge \Lambda^{\ell}|(|a_h|^2 + \cdots + |a_{\ell}|^2)^{1/2}.$$

Let E be the orthogonal complement of F, then E is the vector subspace of $\mathbb{C}^{n+1}$ corresponding to the polar space $A^{\perp}$ of A^{h-1}. Since $a \in E$, by (5.27):

$$\mathop{\mathfrak{M}}_{A^h \supset A^{h-1}} \log|A^h \lrcorner \Lambda^{\ell}|$$
$$= \log|A^{h-1} \lrcorner \Lambda^{\ell}| + \frac{1}{2}\Big(\int_E e^{-\langle Z,Z\rangle} dL\Big)^{-1}$$
$$\cdot \Big(\int_E e^{-\langle Z,Z\rangle} \cdot \log\frac{|z_h|^2+\cdots+|z_{\ell}|^2}{|Z|^2} \cdot dL\Big)$$

where Z is the coordinate function of E, and we write $Z = z_0 e_0 + \cdots + z_n e_n$. The last summand is by definition $a_{h\ell}$. It remains to prove that it is independent of A^{h-1} and Λ^{ℓ} and that it is finite. This amounts to showing that the last integral has these two properties. We first show its independence of A^{h-1} and Λ^{ℓ}. Since each of $e_h,\ldots,e_{\ell}$ is orthogonal to each of $e_0,\ldots,e_{h-1},e_{\ell+1},\ldots,e_n$, each of $e_h,\ldots,e_{\ell}$ is orthogonal to F itself and consequently $\{e_h,\ldots,e_{\ell}\}$ is part of an O.N. basis of E. So we may pick O.N. basis $\{f_h,\ldots,f_n\}$ of E so that $f_h = e_h,\ldots,f_{\ell} = e_{\ell}$. Recall that we wrote $Z = z_0 e_0 + \cdots + z_n e_n$, so that for each β such that $h \le \beta \le \ell$, $z_{\beta} = \langle Z,e_{\beta}\rangle = \langle Z,f_{\beta}\rangle$. Hence if we rewrite the last integral as:

$$\int_{Z\in E} e^{-\langle Z,Z\rangle} \log \frac{|Z,f_h|^2+\cdots+|Z,f_\ell|^2}{|Z|^2} \cdot dL$$

which in turn may be written as

$$\int_{Z\in\mathbb{C}^{n+1-h}} e^{-\langle Z,Z\rangle} \log \frac{|z_0|^2+\cdots+|z_{\ell-h}|^2}{|Z|^2} \cdot dL \ ,$$

where $Z = (z_0,\ldots,z_{n-h})$ is the usual coordinate function on $\mathbb{C}^{n+1-h}$, the independence of the integral from A^{h-1} and Λ^ℓ is obvious. As to finiteness, let us prove that this last integral is convergent. It is equal to

$$\int_{\mathbb{C}^{n+1-h}} e^{-\langle Z,Z\rangle} \log(|z_0|^2+\cdots+|z_{\ell-h}|^2)dL - \int_{\mathbb{C}^{n+1-h}} e^{-\langle Z,Z\rangle} \log|Z|^2 dL$$

Break up each of these integrals into $\int_{|Z|\geq 1} + \int_{|Z|<1}$. The first summand $\int_{|Z|\geq 1}$ in the case of either integral is convergent, as the use of spherical coordinates readily shows. The second summand $\int_{|Z|<1}$ in both cases has a negative integrand and is bounded below by $\int_{|Z|<1} e^{-\langle Z,Z\rangle}\log|z_0|^2 dL$. It suffices to show that this is finite. In turn, it is equivalent to showing that $\int_P e^{-\langle Z,Z\rangle}\log|z_0|^2 dL$ is finite, where P is the unit polycylinder $\{|z_0| < 1,\ldots,|z_{\ell-h}| < 1\}$. But now the integrand is measurable and negative, so we may apply Fubini's theorem and transform the integral into an $(n-h+1)$-fold interated integral each over the unit disc in the complex plane. Therefore, it suffices to show that in $\mathbb{C}$, $\int_{|z_0|<1} e^{-|z_0|^2} \log|z_0|^2 \cdot \frac{\sqrt{-1}}{2}\, dz_0 \wedge d\bar{z}_0$ is finite. By use of polar coordinates, this equals $4\pi\int_0^1 e^{-r^2}(r\log r)dr$, which is certainly finite because the integrand is continuous

Q.E.D.

Armed with Lemma 5.4, we are going to apply the averaging process to (5.26). First, we rewrite it as:

$$(5.28)\qquad (1-\alpha)\int_{r_0}^{r} dt \int_{r_0}^{t} ds \int_{\partial V[s]} \Phi(A^{k-1})*d\tau < CT_k(r) + C'$$

where

$$\Phi(A) = \frac{|A \lrcorner X^{k-1}|^2 |A \lrcorner (b \lrcorner X^{k+1})|^2}{|A \lrcorner X^k|^4} \left(\frac{|A \lrcorner X^k|}{|A \lrcorner (b \lrcorner X^k)|}\right)^{2\alpha}$$

Now fix an A^{k-2} and choose b to be orthogonal to A^{k-2}. We apply $\mathfrak{M}_{A^{k-1}\supset A^{k-2}}$ to both sides of this inequality: the right side remains unchanged while the left side changes to the following:

$$(1-\alpha)\int_{r_0}^{r} dt \int_{r_0}^{t} ds \int_{\partial V[s]} \mathfrak{M}_{A^{k-1}\supset A^{k-2}} \Phi(A^{k-1})$$

(This is just Fubini's theorem and this application is permissible because the integrand is visibly measurable and positive). By Lemma 2.14:

$$\log \mathfrak{M}_{A^{k-1}\supset A^{k-2}} \Phi(A^{k-1}) \geq \mathfrak{M}_{A^{k-1}\supset A^{k-2}} \log \Phi(A^{k-1})$$
$$= \log \Phi(A^{k-2}) + d,$$

where the last step made use of Lemma 5.4 and d is some constant independent of A^{k-2} and dependent only on k. Thus,

$$\mathfrak{M}_{A^{k-1}\supset A^{k-2}} \Phi(A^{k-1}) \geq e^{d}\Phi(A^{k-2})$$

Combining this with (5.28), we obtain:

$$(1-\alpha)\int_{r_0}^{r} dt \int_{r_0}^{t} ds \int_{\partial V[s]} \Phi(A^{k-2}) * d\tau \leq CT_k(r) + C',$$

where C, C' are positive constants independent of α, b and A^{k-2}. We may repeat this process and bring $\Phi(A^{k-2})$ down to $\Phi(A^{k-3})$ etc. The end result is clearly the following: for every integer ℓ such that $0 \leq \ell \leq k$,

$$(1-\alpha)\int_{r_0}^{r} dt \int_{r_0}^{t} ds \int_{\partial V[s]} \Phi(A^{\ell-1}) * d\tau \leq CT_k(r) + C' .$$

where C, C' remain independent of α, b and $A^{\ell-1}$. (If $\ell = 0$, we of course interpret A^{-1} as a 0-vector of unit length, i.e. $e^{\sqrt{-1}\theta}$). In each step, we keep b to be a unit vector orthogonal to the fixed $A^{\ell-1}$. But since we may rewrite $\Phi(A^{\ell-1})$ as

$$\Phi(A^{\ell-1}) = \frac{|A^{\ell-1} \rfloor X^{k-1}|^2 |(A^{\ell-1}\wedge b) \rfloor X^{k+1}|^2}{|A^{\ell-1} \rfloor X^{k}|^4}\left(\frac{|A^{\ell-1} \rfloor X^{k}|}{|(A^{\ell-1}\wedge b) \rfloor X^{k}|}\right)^{2\alpha} ,$$

each factor $A^{\ell-1}\wedge b$ becomes an ℓ-dimensional projective subspace A^{ℓ} of $P_n\mathbb{C}$. Taking this into account, we may summarize the foregoing into

<u>Theorem 5.5</u>. Let $k = 0,\ldots,n-1$ and ℓ be an integer such that $0 \leq \ell \leq k$. Let $A^{\ell-1} \subseteq A^{\ell}$ be projective subspaces of $P_n\mathbb{C}$ of dimensions $\ell-1$ and ℓ respectively ($A^{-1} \equiv e^{\sqrt{-1}\theta}$). Then

$$(1-\alpha)\int_{r_0}^{r} dt \int_{r_0}^{t} ds \int_{\partial V[s]} \frac{|A^{\ell-1} \lrcorner X^{k-1}|^2 |A^{\ell} \lrcorner X^{k+1}|^2}{|A^{\ell-1} \lrcorner X^{k}|^4} \cdot \left(\frac{|A^{\ell-1} \lrcorner X^{k}|}{|A^{\ell} \lrcorner X^{k}|}\right)^{2\alpha} {}_*d\tau$$

$$\leq CT_k(r) + C'$$

where C, C' are positive constants independent of α, $A^{\ell-1}$ and A^{ℓ}, while $0 < \alpha < 1$. (X^k etc. $= X^k_\sigma$ etc.)

There is a similar inequality for $\ell > k$. Its derivation is dual to the above, so we will only sketch it. We consider the contracted curve of the second kind $X^k \lrcorner A^{k+1}: V \to A^{k+1}$ given by (3.11) of Chapter III, §6. We apply (5.24) to the holomorphic curve $X^k \lrcorner A^{k+1}$, so $\tilde{x}$ will be replaced by $(X^k \lrcorner A^{k+1})$ and $\tilde{x} \wedge \tilde{x}^{(1)}$ by $\langle X^{k+1}, A^{k+1}\rangle (X^{k-1} \lrcorner A^{k+1})$ (Lemma 3.11). The order function of $X^k \lrcorner A^{k+1}$ is simply $T_k(r, A^{k+1})$, by (4.34). However, so far as the inequality (5.24) is concerned, Lemma 4.11 says that $T_k(r, A^{k+1})$ may be replaced by $T_k(r) + b_k$, where b_k is a constant independent of A^{k+1}. Putting all these together, we obtain

$$(5.29)\quad (1-\alpha)\int_{r_0}^{r} dt \int_{r_0}^{t} ds \int_{\partial V[s]} \frac{|X^{k+1}, A^{k+1}|^2 |b \lrcorner (X^{k-1} \lrcorner A^{k+1})|^2}{|X^{k} \lrcorner A^{k+1}|^4} \cdot \left(\frac{|X^{k} \lrcorner A^{k+1}|}{|b, X^{k} \lrcorner A^{k+1}|}\right)^{2\alpha} {}_*d\tau$$

$$\leq CT_k(r) + C'$$

where $0 < \alpha < 1$, and C, C' are positive constants independent of α, b and A^{k+1}.

We are going to subject (5.29) to a similar kind of averaging process as (5.27). Let $\Phi(A^h)$ be a function of projective subspaces of dimension h in $P_n\mathbb{C}$. Fix an A^{h+1} and consider the set of $A^h \subseteq A^{h+1}$. The polar space of such an A^h in A^{h+1} is just a point $a \in A^{h+1}$, and conversely each $a \in A^{h+1}$ also uniquely determines an $A^h \subseteq A^{h+1}$: A^h is the polar space of a in A^{h+1}. If we restrict the domain of definition of Φ to only such A^h's contained in the fixed A^{h+1}, Φ therefore becomes a function of $a \in A^{h+1}$, which we simply denote by $\Phi(a)$. By definition, <u>the average</u> $\mathfrak{M}_{A^h \subseteq A^{h+1}} \Phi(A^h)$ <u>of</u> Φ <u>over all</u> A^h <u>contained in the fixed</u> A^{h+1} is the arithmetic mean of $\Phi(a)$ over A^{h+1}, i.e.,

$$\mathfrak{M}_{A^h \subseteq A^{h+1}} \Phi(A^h) = \left(\int_{A^{h+1}} \Omega\right)^{-1}\left(\int_{A^{h+1}} \Phi(a)\Omega(a)\right)$$

where Ω is the volume form of the F-S metric on A^{h+1}. Let E be the vector subspace of $\mathbb{C}^{n+1}$ corresponding to A^{h+1}, then Theorem 5.2 says that we may equivalently define:

$$\mathfrak{M}_{A^h \subseteq A^{h+1}} \Phi(A^h) = \left(\int_E e^{-\langle Z,Z\rangle} dL\right)^{-1}\left(\int_E e^{-\langle Z,Z\rangle} \Phi\left(\left[\frac{Z}{|Z|}\right]\right) dL\right)$$

where dL denotes the Lebesgue measure of E, Z its usual coordinate function and $\left[\frac{Z}{|Z|}\right]$ the point of A^{h+1} represented by $\frac{Z}{|Z|}$. The following analogue of Lemma 5.4 may be proved in a similar manner.

<u>Lemma 5.6</u>. Let Λ^ℓ be a fixed decomposable $(\ell+1)$-vector and let h be an integer such that $\ell \leq h \leq n-1$. Then for a

fixed A^{h+1},

$$\underset{A^h \subseteq A^{h+1}}{\mathfrak{M}} \log|\Lambda^\ell \lrcorner A^h| = \log|\Lambda^\ell \lrcorner A^{h+1}| + a_{\ell h},$$

where $a_{\ell h}$ is a constant depending only on ℓ and h, and independent of Λ^ℓ and A^{h+1}.

Rewrite (5.29) as:

$$(1-\alpha)\int_{r_o}^{r} dt \int_{r_o}^{t} ds \int_{\partial V[s]} \Phi(A^{k+1})*d\tau \leq CT_k(r) + C',$$

where

$$\Phi(A) = \frac{|(b\wedge X^{k-1})\lrcorner A|^2 |X^{k+1}\lrcorner A|^2}{|X^k \lrcorner A|^4} \left(\frac{|X^k \lrcorner A|}{|(b\wedge X^k)\lrcorner A|}\right)^{2\alpha}.$$

Now hold b fixed and take the average of both sides of this inequality over all A^{k+1} contained in a fixed A^{k+2}. Making use of $\log(\mathfrak{M}\,\Phi) \geq \mathfrak{M}(\log \Phi)$ and Lemma 5.6 as above, we arrive at:

$$(1-\alpha)\int_{r_o}^{r} dt \int_{r_o}^{t} ds \int_{\partial V[s]} \Phi(A^{k+2})*d\tau \leq CT_k(r) + C'$$

where C, C' remain independent of α, b and A^{k+2}. Repeated application of this process leads us to the following: for every integer ℓ such that $k \leq \ell \leq n-1$,

$$(1-\alpha)\int_{r_o}^{r} dt \int_{r_o}^{t} ds \int_{\partial V[s]} \Phi(A^{\ell+1})*d\tau \leq CT_k(r) + C'$$

where C, C' are independent of α, b and $A^{\ell+1}$. Let us rewrite $\Phi(A^\ell)$ in this form:

$$\Phi(A^{\ell+1}) = \frac{|X^{k-1}\lrcorner(b\lrcorner A^{\ell+1})|^2 |X^{k+1}\lrcorner A^{\ell+1}|^2}{|X^k \lrcorner A^{\ell+1}|^4} \left(\frac{|X^k \lrcorner A^{\ell+1}|}{|X^k \lrcorner (b\lrcorner A^{\ell+1})|}\right)^{2\alpha}$$

What remains to be done is clearly to choose b cleverly so that the factor $b \lrcorner A^{\ell+1}$ becomes meaningful. We choose b to be a point of $A^{\ell+1}$, then $b \lrcorner A^{\ell+1}$ is an ℓ-dimensional projective subspace of $A^{\ell+1}$, in fact, the polar space of b in $A^{\ell+1}$. With this choice, we have arrived at the following counterpart of Theorem 5.5.

Theorem 5.7. Let $k = 0,\ldots,n-1$ and ℓ be an integer such that $k \leq \ell \leq n-1$. Let $A^{\ell} \subseteq A^{\ell+1}$ be projective subspaces in $P_n\mathbb{C}$ of dimensions ℓ and $\ell+1$ respectively. Then

$$(1-\alpha)\int_{r_0}^{r} dt \int_{r_0}^{t} ds \int_{\partial V[s]} \frac{|X^{k-1} \lrcorner A^{\ell}|^2 |X^{k+1} \lrcorner A^{\ell+1}|^2}{|X^k \lrcorner A^{\ell+1}|^4} \cdot \left(\frac{|X^k \lrcorner A^{\ell+1}|}{|X^k \lrcorner A^{\ell}|}\right)^{2\alpha} {*d\tau}$$

$$\leq CT_k(r) + C'$$

where $0 < \alpha < 1$ and C, C' are positive constants independent of α, A^{ℓ} and $A^{\ell+1}$.

§4. We now head towards the defect relations. Throughout this entire section, we will follow closely Ahlfors [1], p. 20ff. This part of Ahlfors' paper constitutes, in my opinion, one of the most brilliant chapters in the annals of mathematics.

We must first discuss the notion of a system of h-dimensional projective spaces $\{A^h\}$ in general position. Now, $\{A^h\} \subseteq G(n,h) \subseteq P_{\ell(h)-1}\mathbb{C}$, where $\ell(h) = \binom{n+1}{h+1}$. The $\{A^h\}$

correspond to points in $P_{\ell(h)-1}\mathbb{C}$ and hence to a system of one-dimensional subspaces of $\mathbb{C}^{\ell(h)}$ via the projection map of the fibration ${}_h\pi\colon \mathbb{C}^{\ell(h)} - \{0\} \to P_{\ell(h)-1}\mathbb{C}$. In general, given any j one-dimensional subspaces of $\mathbb{C}^{\ell(h)}$, there is always a j-dimensional subspace of $\mathbb{C}^{\ell(h)}$ which contains them all, e.g. any j-dimensional subspace that contains their common span. We now adopt as formal definition that a system of h-dimensional projective subspaces $\{A^h\}$ of $P_n\mathbb{C}$ is said to be <u>in general position</u> if and only if, when considered as one-dimensional subspaces of $\mathbb{C}^{\ell(h)}$, each j-dimensional subspace of $\mathbb{C}^{\ell(h)}$ contains no more than j of the $\{A^h\}$ for all j such that $1 \leq j \leq \ell(h)-1$. The following lemma somewhat clarifies this definition.

<u>Lemma 5.8</u>. The following are equivalent conditions on a system of h-dimensional projective subspaces $\{A^h\}$ of $P_n\mathbb{C}$:

(a) The $\{A^h\}$ are in general position.

(b) Considered as one-dimensional subspaces of $\mathbb{C}^{\ell(h)}$, any j of the $\{A^h\}$ are linearly independent, for $1 \leq j \leq \ell(h)$.

(c) Considered as one-dimensional subspaces of $\mathbb{C}^{\ell(h)}$, any j of the $\{A^h\}$ span a j-dimensional subspace of $\mathbb{C}^{\ell(h)}$, for $1 \leq j \leq \ell(h)$.

Any of the above implies that

(d) The intersection of any i of the $\{A^h\}$ is a projective subspace of $P_n\mathbb{C}$ of dimension at most $h-i+1$, for $1 \leq i \leq h+2$. (By definition, a projective subspace of dimension -1 is the empty set.)

In general, (d) does not imply any of (a)-(c), but it does in a special case. For a later reference, we single this out.

Lemma 5.9. A system of hyperplanes $\{A^{n-1}\}$ of $P_n\mathbb{C}$ is in general position if and only if the intersection of any i of them is a projective subspace of dimension $n-i$ in $P_n\mathbb{C}$ $(1 \leq i \leq n+1)$.

The proofs of all these statements are simple exercises in linear algebra and so will be left to the reader. The following is a bit deeper.

Lemma 5.10. Let B^k be a given k-dimensional projective subspace of $P_n\mathbb{C}$. For a system of h-dimensional projective subspaces $\{A^h\}$ in general position, the number of A^h with the property that for a fixed ℓ $(\ell \leq h,\ \ell \leq k)$, every $A^\ell \subseteq A^h$ satisfies $A^\ell \lrcorner B^k = 0$, does not exceed

$$p_h(k,\ell) \equiv \binom{n+1}{h+1} - \Sigma_{j\geq 0} \binom{k+1}{\ell+j+1}\binom{n-k}{h-\ell-j},$$

where the binomial coefficients $\binom{\mu}{\nu}$ are defined for all integers by the expansion $(1+x)^\mu = \Sigma_{\nu=-\infty}^{\nu=\infty} \binom{\mu}{\nu} x^\nu$.

Proof. Choose O.N. basis $e_0,\dots,e_n$ of $\mathbb{C}^n$ so that $B = e_0 \wedge \cdots \wedge e_k$. Let $A^h = \Sigma\, a_{i_0\cdots i_h} e_{i_0} \wedge \cdots \wedge e_{i_h}$. Then $A^\ell \lrcorner B^k = 0$ for all $A^\ell \subseteq A^h$ if and only if the coefficients of A^h satisfy the condition:

(5.30) $\quad a_{i_0\cdots i_h} = 0$ whenever at least $(\ell+1)$ of the subscripts are smaller than or equal to k.

This is obvious. Now for $j \geq 0$, the number of components which have exactly $(\ell+1)+j$ subscripts smaller than or equal to k is $\binom{k+1}{(\ell+1)+j}\binom{n-k}{(h+1)-\{(\ell+1)+j\}} = \binom{k+1}{\ell+j+1}\binom{n-k}{h-\ell-j}$. So the total number of coefficients satisfying the condition stipulated in (5.30) is

$$\overline{p}_h(k,\ell) = \Sigma_{j\geq 0}\binom{k+1}{\ell+j+1}\binom{n-k}{h-\ell-j} .$$

Let $\{\lambda_{i_o \cdots i_h} : i_o < \cdots < i_h\}$ be the usual coordinate functions on $\mathbb{C}^{\ell(k)}$. Consider the subspace S of $\mathbb{C}^{\ell(h)}$ defined by:

$$\begin{cases} \lambda_{i_o \cdots i_h} = 0 \text{ for all } i_o, \ldots, i_h \text{ where at least } (\ell+1) \\ \text{of them are smaller than or equal to } k. \end{cases}$$

We have just seen that there are exactly $\overline{p}_h(k,\ell)$ such (obviously independent) linear equations, so $\dim S = \ell(h) - \overline{p}_h(k,\ell)$. We have also seen that an A^h has the property that every $A^\ell \subseteq A^h$ satisfies $A^\ell \lrcorner B^k = 0$ if and only if A^h lies in S. Since $\{A^h\}$ is a system in general position, it follows that the number of such A^h satisfying $A^\ell \lrcorner B^k = 0$ for every $A^\ell \subseteq A^h$ cannot exceed the dimension of S, which is $\ell(h) - \overline{p}_h(k,\ell)$. Q.E.D.

Now let there be given a projective subspace A^h in $P_n\mathbb{C}$ of dimension h. Suppose $A^h = e_o \wedge \cdots \wedge e_h$, where $e_o, \ldots, e_h$ are O.N. and will be fixed throughout the remaining discussion. For a fixed ℓ such that $0 \leq \ell \leq h$, let $A^\ell = e_{i_o} \wedge \cdots \wedge e_{i_\ell}$ where $0 \leq i_o < \cdots < i_\ell \leq h$, and let $A^{\ell-1} \subseteq A^\ell$ be such that $A^{\ell-1} = e_{j_0} \wedge \cdots \wedge e_{j_{\ell-1}}$, $j_0 < \cdots < j_{\ell-1}$. (Recall that A^{-1} is just a scalar of absolute value one). We apply Theorem 5.5 to this pair $A^{\ell-1} \subseteq A^{\ell}$, $0 \leq \ell \leq k$, thereby obtaining:

$$(1-\alpha)\int_{r_0}^{r} dt \int_{r_0}^{t} ds \int_{\partial V[s]} \frac{|A^{\ell-1} \lrcorner X^{k-1}|^2 |A^{\ell} \lrcorner X^{k+1}|^2}{|A^{\ell-1} \lrcorner X^{k}|^4} \cdot \left(\frac{|A^{\ell-1} \lrcorner X^{k}|}{|A^{\ell} \lrcorner X^{k}|}\right)^{2\alpha} * d\tau$$
$$\leq CT_k(r) + C',$$

where $0 < \alpha < 1$, and C, C' are positive constants independent of α, $A^{\ell-1}$ and A^{ℓ}. We may rewrite the integrand as:

$$\frac{|A^{\ell-1} \lrcorner X^{k-1}|^2}{|A^{\ell-1} \lrcorner X^{k}|^{4-2\alpha}} \cdot \frac{|A^{\ell} \lrcorner X^{k+1}|^2}{|A^{\ell} \lrcorner X^{k}|^{2\alpha}}$$

Now for each choice of $i_0 < \cdots < i_\ell$ and $j_0 < \cdots < j_{\ell-1}$, subject only to the conditions that $0 \leq i_0, \ldots, i_\ell \leq h$ and $A^{\ell-1} \subseteq A^{\ell}$, we have an inequality as above. If we now let $i_0, \ldots, i_\ell$ and $j_0, \ldots, j_{\ell-1}$ take on all possible integral values and sum all of the resulting inequalities, we get

$$(1-\alpha)\int_{r_0}^{r} dt \int_{r_0}^{t} ds \int_{\partial V[s]} \Sigma \frac{|A^{\ell-1} \lrcorner X^{k-1}|^2}{|A^{\ell-1} \lrcorner X^{k}|^{4-2\alpha}} \cdot \frac{|A^{\ell} \lrcorner X^{k+1}|^2}{|A^{\ell} \lrcorner X^{k}|^{2\alpha}} * d\tau$$
$$\leq CT_k(r) + C'$$

with new values of C and C' which, however, remain independent of α and A^h. The finite sum in the integrand is greater than every single term, and clearly there is a term which is at least equal to

$$\operatorname{Max} \frac{|A^{\ell-1} \lrcorner X^{k-1}|^2}{|A^{\ell-1} \lrcorner X^k|^{4-2\alpha}} \cdot \operatorname{Min} \frac{|A^{\ell} \lrcorner X^{k+1}|^2}{|A^{\ell} \lrcorner X^k|^{2\alpha}} ,$$

where the maximum and minimum refer to the finite number of $A^{\ell-1}$ and A^{ℓ} over which we have summed. The expression can be written in the equivalent form:

$$\frac{|X^{k-1}|^2 |X^{k+1}|^2}{|X^k|^4} \frac{\operatorname{Min}\left(\frac{|X^k|^{2\alpha}}{|A^{\ell} \lrcorner X^k|^{2\alpha}} : \frac{|X^{k+1}|^2}{|A^{\ell} \lrcorner X^{k+1}|^2}\right)}{\operatorname{Min}\left(\frac{|X^{k-1}|^2}{|A^{\ell-1} \lrcorner X^{k-1}|^2} : \frac{|X^k|^{4-2\alpha}}{|A^{\ell-1} \lrcorner X^k|^{4-2\alpha}}\right)}$$

Now $\frac{|X^k|^{4-2\alpha}}{|A^{\ell-1} \lrcorner X^k|^{4-2\alpha}} > \frac{|X^k|^2}{|A^{\ell-1} \lrcorner X^k|^2}$ because $4-2\alpha > 2$ and because $\frac{|X^k|}{|A^{\ell-1} \lrcorner X^k|} \geq 1$ (by (1.12)). Having done this, we obtain:

$$(1-\alpha) \int_{r_0}^{r} dt \int_{r_0}^{t} ds \int_{\partial V[s]} \frac{|X^{k-1}|^2 |X^{k+1}|^2}{|X^k|^4}$$
$$\cdot \frac{\operatorname{Min}\left(\frac{|X^k|^{2\alpha}}{|A^{\ell} \lrcorner X^k|^{2\alpha}} : \frac{|X^{k+1}|^2}{|A^{\ell} \lrcorner X^{k+1}|^2}\right)}{\operatorname{Min}\left(\frac{|X^{k-1}|^2}{|A^{\ell-1} \lrcorner X^{k-1}|^2} : \frac{|X^k|^2}{|A^{\ell-1} \lrcorner X^k|^2}\right)} * d\tau$$
$$\leq CT_k(r) + C'$$

For simplicity, let us denote the quotient of the two minima by $\Phi(A^h)$. We now let A^h range over a finite system in general position and form the sum of the corresponding inequalities

$$(5.31) \quad (1-\alpha) \int_{r_0}^{r} dt \int_{r_0}^{t} ds \int_{\partial V[s]} \frac{|X^{k-1}|^2 |X^{k+1}|^2}{|X^k|^4} \sum_{A^h} \Phi(A^h) * d\tau$$
$$\leq CT_k(r) + C'$$

with new constants C, C' which are still independent of α.

By Lemma 5.10, it is impossible to have $\operatorname{Min} \frac{|X^k|}{|A^\ell \lrcorner X^k|} = \infty$ for more than $p_h(k,\ell)$ of the given A^h. Therefore, there exists an M such that $\operatorname{Min} \frac{|X^k|}{|A^\ell \lrcorner X^k|} \geq M$ for at most $p_h(k,\ell)$ of the A^h, independently of X^k. The remaining A^h's satisfy

$$\operatorname{Min} \frac{|X^k|}{|A^\ell \lrcorner X^k|} \leq M.$$

Let us say there are p such A^h's. Because $\frac{|X^{k+1}|}{|A^\ell \lrcorner X^{k+1}|} \geq 1$ (by (1.12)), we have for these p A^h's:

$$\operatorname{Min}\left(\frac{|X^k|^{2\alpha}}{|A^\ell \lrcorner X^k|^{2\alpha}} : \frac{|X^{k+1}|^2}{|A^\ell \lrcorner X^{k+1}|^2}\right) \leq M^{2\alpha} .$$

Furthermore,

$$\frac{|X^{k-1}|^2}{|A^{\ell-1} \lrcorner X^{k-1}|^2} : \frac{|X^k|^2}{|A^{\ell-1} \lrcorner X^k|^2} = \left(\frac{\left|A^{\ell-1} \lrcorner \frac{X^k}{|X^k|}\right|}{\left|A^{\ell-1} \lrcorner \frac{X^{k-1}}{|X^{k-1}|}\right|}\right)^2 \geq 1$$

in view of (1.13). Therefore its minimum will still exceed one. Consequently, referring back to the expression of $\Phi(A^h)$, we get

$$\Phi(A^h) \leq M^{2\alpha},$$

for these p of the A^h's. Let there be q of the A^h's such that $\operatorname{Min} \frac{|X^k|}{|A^\ell \lrcorner X^k|} \geq M$. By the above, $q \leq p_h(k,\ell)$. So, if we use Σ' to denote summation over these q of the A^h's, we have:

$$\begin{aligned}\log \sum_{A^h} \Phi(A^h) &> \log \Sigma' \ \Phi(A^h) \\ &= \log\{\frac{1}{q}\Sigma' \ \Phi(A^h)\} + \log q \\ &\geq \frac{1}{q}\Sigma' \log \Phi(A^h) + \log q \quad \text{(Lemma 2.14)} \\ &\geq \frac{1}{q}\{\sum_{A^h} \log \Phi(A^h) - 2\alpha p \log M\} + \log q \\ &\geq \frac{1}{p_h(k,\ell)}\{\sum_{A^h} \log \Phi(A^h) - 2\alpha p \log M\} + \log q,\end{aligned}$$

i.e. $\log \sum_{A^h} \Phi(A^h) \geq \frac{1}{p_h(k,\ell)} \sum_{A^h} \log \Phi(A^h) + c_1$ where c_1 is a constant depending only on the system $\{A^h\}$. Let L be as usual the constant $\int_{\partial V[s]} *d\tau$ for $s \geq r(\tau)$. By the Concavity of the Logarithm again, we have:

$$\begin{aligned}&\log\{\frac{1}{L}\int_{\partial V[s]} \frac{|X^{k-1}|^2|X^{k+1}|^2}{|X^k|^2} \sum_{A^h} \Phi(A^h)*d\tau\} \\ &\geq \frac{1}{L}\int_{\partial V[s]} \log\{\frac{|X^{k-1}|^2|X^{k+1}|^2}{|X^k|^4} \cdot \sum_{A^h} \Phi(A^h)\}*d\tau \\ &= \frac{1}{L}\int_{\partial V[s]} \log\frac{|X^{k-1}|^2|X^{k+1}|^2}{|X^k|^4} *d\tau + \frac{1}{L}\int_{\partial V[s]} \log \sum_{A^h} \Phi(A^h)*d\tau \\ &= \frac{2\pi}{L}\{E(s) + S_k(s) + (T_{k-1}(s) - 2T_k(s) + T_{k+1}(s)) \\ &\qquad + \int_{\partial V[r_o]} \log \frac{|X^{k-1}|^2|X^{k+1}|^2}{|X^k|^4} *d\tau\} + \frac{1}{L}\int_{\partial V[s]} \log \sum_{A^h} \Phi(A^h)*d\tau \\ &\qquad\qquad \text{(by Theorem 4.24)} \\ &\geq \frac{2\pi}{L}\{E(s) + S_k(s) + (T_{k-1}(s) - 2T_k(s) + T_{k+1}(s))\} \\ &\qquad + \frac{1}{Lp_h(k,\ell)} \sum_{A^h} \int_{\partial V[s]} \log \Phi(A^h)*d\tau + c_2\end{aligned}$$

where c_2 is a constant depending only on the system $\{A^h\}$. Keeping this inequality in mind, we inspect the integral $\int_{\partial V[s]} \log \Phi(A^h) * d\tau$. It is equal to

$$\int_{\partial V[s]} \log \operatorname{Min}\left(\frac{|X^k|^{2\alpha}}{|A^\ell \lrcorner X^k|^{2\alpha}} : \frac{|X^{k+1}|^2}{|A^\ell \lrcorner X^{k+1}|}\right) * d\tau$$

$$- \int_{\partial V[s]} \log \operatorname{Min}\left(\frac{|X^{k-1}|^2}{|A^{\ell-1} \lrcorner X^{k-1}|^2} : \frac{|X^k|^2}{|A^{\ell-1} \lrcorner X^k|^2}\right) * d\tau$$

$$\overset{\text{def}}{\equiv} \int_{\partial V[s]} I * d\tau - \int_{\partial V[s]} II * d\tau$$

Obviously,

$$I \geq \log \operatorname{Min}\left(\frac{|X^k|^2}{|A^\ell \lrcorner X^k|^2} : \frac{|X^{k+1}|^2}{|A^\ell \lrcorner X^{k+1}|^2}\right)$$

$$- (1-\alpha) \log \operatorname{Max} \frac{|X^k|^2}{|A^\ell \lrcorner X^k|^2}$$

$$\geq \log \operatorname{Min}\left(\frac{|X^k|^2}{|A^\ell \lrcorner X^k|^2} : \frac{|X^{k+1}|^2}{|A^\ell \lrcorner X^{k+1}|^2}\right)$$

$$- (1-\alpha) \sum_{A^\ell \subseteq A^h} \log \frac{|X^k|^2}{|A^\ell \lrcorner X^k|^2} .$$

Now by Lemma 4.10,

$$\frac{1}{2\pi} \int_{\partial V[s]} \log \frac{|X^k|}{|A^\ell \lrcorner X^k|} * d\tau \leq T_k(s) + a_k$$

where a_k is independent A^ℓ and s. Therefore,

$$\int_{\partial V[s]} I * d\tau \geq \int_{\partial V[s]} \log \operatorname{Min}\left(\frac{|X^k|^2}{|A^\ell \lrcorner X^k|^2} : \frac{|X^{k+1}|^2}{|A^\ell \lrcorner X^{k+1}|^2}\right) * d\tau$$

$$- (1-\alpha)\, c_3\, (T_k(s) + a_k)$$

where c_3 and a_k are constants depending only on the system

$\{A^h\}$ alone and not on α. We now choose α: α should satisfy $(1-\alpha)T_k(s) = 1$. Hence there is a new constant c_4 depending only on $\{A^h\}$ such that

$$\int_{\partial V[s]} I * d\tau \geq \int_{\partial V[s]} \log \operatorname{Min}\left(\frac{|X^k|^2}{|A^\ell \lrcorner X^k|^2} : \frac{|X^{k+1}|^2}{|A^\ell \lrcorner X^{k+1}|^2}\right) * d\tau + c_4$$

and therefore

$$\begin{aligned}
&\int_{\partial V[s]} \log \Phi(A^h) * d\tau \\
&\quad \geq \int_{\partial V[s]} \log \operatorname{Min}\left(\frac{|X^k|^2}{|A^\ell \lrcorner X^k|^2} : \frac{|X^{k+1}|^2}{|A^\ell \lrcorner X^{k+1}|^2}\right) * d\tau \\
&\qquad - \int_{\partial V[s]} \log \operatorname{Min}\left(\frac{|X^{k-1}|^2}{|A^{\ell-1} \lrcorner X^{k-1}|^2} : \frac{|X^k|^2}{|A^{\ell-1} \lrcorner X^k|^2}\right) * d\tau + c_4,
\end{aligned}$$

so that taking into account of a previous inequality, we obtain:

$$\begin{aligned}
&\log\left\{\frac{1}{L} \int_{\partial V[s]} \frac{|X^{k-1}|^2 |X^{k+1}|^2}{|X^k|^2} \sum_{A^h} \Phi(A^h) * d\tau\right\} \\
&\quad \geq \frac{2\pi}{L}\{E(s) + S_k(s) + (T_{k-1}(s) - 2T_k(s) + T_{k+1}(s))\} \\
&\qquad + \frac{1}{Lp_h(k,\ell)} \sum_{A^h} \left\{ \int_{\partial V[s]} \log \operatorname{Min}\left(\frac{|X^k|^2}{|A^\ell \lrcorner X^k|^2} : \frac{|X^{k+1}|^2}{|A^\ell \lrcorner X^{k+1}|^2}\right) * d\tau \right. \\
&\qquad\qquad \left. - \int_{\partial V[s]} \log \operatorname{Min}\left(\frac{|X^{k-1}|^2}{|A^{\ell-1} \lrcorner X^{k-1}|^2} : \frac{|X^k|^2}{|A^{\ell-1} \lrcorner X^k|^2}\right) * d\tau \right\} \\
&\qquad + c_5
\end{aligned}$$

where c_5 is a constant depending only on the system $\{A^h\}$.

Now define:

$$\theta(s) = p_h(k,\ell)\{E(s) + S_k(s) + (T_{k-1}(s) - 2T_k(s) + T_{k+1}(s))\}$$

$$+ \sum_{A^h} \frac{1}{2\pi} \int_{\partial V[s]} \log \operatorname{Min}\left(\frac{|X^k|^2}{|A^\ell \lrcorner X^k|^2} : \frac{|X^{k+1}|^2}{|A^\ell \lrcorner X^{k+1}|^2}\right) * d\tau$$

$$- \sum_{A^h} \frac{1}{2\pi} \int_{\partial V[s]} \log \operatorname{Min}\left(\frac{|X^{k-1}|^2}{|A^{\ell-1} \lrcorner X^{k-1}|^2} : \frac{|X^k|^2}{|A^{\ell-1} \lrcorner X^k|^2}\right) * d\tau$$

Then the above may be rewritten as:

$$\exp\left\{\frac{2\pi}{Lp_h(k,\ell)}\,\theta(s) + c_5\right\} \leq \int_{\partial V[s]} \frac{|X^{k-1}|^2 |X^{k+1}|^2}{|X^k|^2} \sum_{A^h} \Phi(A^h) * d\tau$$

By (5.31), this implies

$$(1-\alpha)\int_{r_o}^{r} dt \int_{r_o}^{t} \exp\left\{\frac{2\pi}{Lp_h(k,\ell)}\,\theta(s) + c_5\right\} ds \leq CT_k(r) + C'.$$

Recall that we have already chosen α so that $(1-\alpha)T_k(r) = 1$. Hence:

$$\int_{r_o}^{r} dt \int_{r_o}^{t} \exp\left\{\frac{2\pi}{Lp_h(k,\ell)}\,\theta(s) + c_5\right\} ds \leq CT_k^2(r) + C'T_k(r)$$
$$< C''T_k^2(r)$$

where C'' is some new constant and the last inequality is because $T_k(r)$ is monotone increasing. In the notation of Chapter IV, §7, this may be written as:

$$\theta + c_5 = \mu(T_k^2)$$

or in view of Lemma 4.16(ii),

$$\theta = \mu(T_k^2).$$

Recall that we have defined a function $T(r) = \max\{T_o(r),\ldots,T_{n-1}(r)$

So by Lemma 4.16(i), we obtain

$$\theta = \mu(T^2).$$

In greater detail, we have the following:

$$(5.32)\quad \sum_{A^h} \frac{1}{2\pi} \int_{\partial V[s]} \log \mathrm{Min}\Big(\frac{|X^k|^2}{|A^\ell \lrcorner X^k|^2} : \frac{|X^{k+1}|^2}{|A^\ell \lrcorner X^{k+1}|^2}\Big) * d\tau$$

$$- \sum_{A^h} \frac{1}{2\pi} \int_{\partial V[s]} \log \mathrm{Min}\Big(\frac{|X^{k-1}|^2}{|A^{\ell-1} \lrcorner X^k|^2} : \frac{|X^k|^2}{|A^{\ell-1} \lrcorner X^k|^2}\Big) * d$$

$$= p_h(k,\ell)(-E - S_k - (T_{k-1} - 2T_k + T_{k+1})) + \mu(T^2).$$

We wish to point out explicitly that (5.32) is only valid for $0 \leq \ell \leq k$. Introduce the notation:

$$\Psi(k,\ell) = \sum_{A^h} \frac{1}{2\pi} \int_{\partial V[s]} \log \mathrm{Min}\Big(\frac{|X^k|^2}{|A^\ell \lrcorner X^k|^2} : \frac{|X^{k+1}|^2}{|A^\ell \lrcorner X^{k+1}|^2}\Big) * d\tau$$

Then (5.32) may be written as:

$$(5.32)^k_\ell \quad \Psi(k,\ell) - \Psi(k-1,\ell-1) = p_h(k,\ell)(-E - S_k - T_{k-1} + 2T_k - T_{k+1}) + \mu(T^2)$$

In a similar fashion, we obtain:

$$(5.32)^{k-1}_{\ell-1} \quad \Psi(k-1,\ell-1) - \Psi(k-2,\ell-2) = p_h(k-1,\ell-1)(-E - S_k - T_{k-2} + 2T_{k-1} - T_k) + \mu(T^2)$$

. . . .

$$(5.32)^{k-\ell-1}_1 \quad \Psi(k-\ell+1,1) - \Psi(k-\ell,0) = p_h(k-\ell+1,1)(-E - S_{k-\ell+1} - T_{k-\ell} + 2T_{k-\ell+1} - T_{k-\ell-2}) + \mu(T^2)$$

$(5.32)_0^{k-\ell}$ $\Psi(k-\ell,0)$

$$= p_h(k-\ell,0)(-E - S_{k-\ell} - T_{k-\ell-1} + 2T_{k-\ell} - T_{k-\ell+1}) + \mu(T^2)$$

The last is because $\Psi(k-\ell-1,-1) = 0$ (This explains why we have been so careful all along about the case A^{-1}). Adding all these inequalities $(5.32)_\ell^k,\ldots,(5.32)_0^{k-\ell}$, we obviously get:

$$\sum_{A^h} \frac{1}{2\pi}\int_{\partial V[s]} \log \operatorname{Min}\left(\frac{|X^k|^2}{|A^\ell \lrcorner X^k|^2} : \frac{|X^{k+1}|^2}{|A^\ell \lrcorner X^{k+1}|^2}\right) * d\tau$$

$$= \Sigma_{i=0}^{\ell} p_h(k-i,\ell-i)(-E - S_{k-i} - T_{k-i-1} + 2T_{k-i} - T_{k-i+1}) + \mu(T^2)$$

where use has been made of Lemma 4.16(v). Finally, we let $\ell = h$. In this case, there is only one $A^\ell \subset A^h$ and no minimum need be taken. Recall at this point the definition of $m_k(r,A^h)$ in (4.29):

$$m_k(r,A^h) = \frac{1}{2\pi}\int_{\partial V[t]} \log \frac{|X^k|}{|A^h \lrcorner X^k|} * d\tau \Big|_{r_0}^{r}$$

So the left hand side of the above becomes:

$$\sum_{A^h} \{m_k(r,A^h) - m_{k+1}(r,A^h)\} + \frac{1}{2\pi}\int_{\partial V[r_0]} \log\frac{|X^k|}{|A^h \lrcorner X^k|} * d\tau$$
$$- \frac{1}{2\pi}\int_{\partial V[r_0]} \log\frac{|X^{k+1}|}{|A^h \lrcorner X^{k+1}|} * d\tau$$

By (the analogue of) Lemma 4.6, the last two terms are continuous functions of A^h and so both are $O(1)$ as $r \to s$ (s as in Definition 2.1). We may therefore apply Lemma 4.16(ii) to conclude:

$$\sum_{A^h} \{m_k(r,A^h) - m_{k+1}(r,A^h)\}$$
$$= \Sigma_{i=0}^{h} p_h(k-i,h-i)(-E - S_{k-i} - T_{k-i-1} + 2T_{k-i} - T_{k-i+1})$$
$$+ \mu(T^2)$$

This is valid for $0 \le h \le k$. Note that $m_n(r,A^h) \equiv 0$, so upon summing over k from k to $n-1$ and writing $m_k(A^h)$ for $m_k(r,A^h)$, we finally arrive at:

$$(5.33)\quad \sum_{A^h} m_k(A^h) = \Sigma_{m=k}^{n-1} \Sigma_{i=0}^{h} p_h(m-i,h-i)(-E - S_{m-i})$$
$$+ \Sigma_{m=k}^{n-1} \Sigma_{i=0}^{h} p_h(m-i,h-i)(-T_{m-i-1} + 2T_{m-i}$$
$$- T_{m-i+1}) + \mu(T^2)$$

Once more, we emphasize that (5.33) is valid only for $0 \le h \le k$. This is essentially one-half of the sought for defect relations. In all applications, we are only interested in the case $h = k$, but there is as yet no direct method of proving (5.33) only for this case.

Our next objective is to apply some combinatorics to simplify the last double sum of (5.33). We will extend this last double sum to all i, $-\infty < i < \infty$, taking $T_i = 0$ for $i < 0$ and for $i > n-1$. Recall that

$$p_h(k,\ell) = \binom{n+1}{h+1} - \Sigma_{j\geq 0}\binom{k+1}{\ell+j+1}\binom{n-k}{h-\ell-j},$$

where $\binom{\mu}{\nu}$ is defined for all integers by the binomial series $(1+x)^{\mu} = \Sigma_{\nu=-\infty}^{\nu=+\infty} \binom{\mu}{\nu}x^{\nu}$. (Note that $\binom{\mu}{\nu} = 0$ if $\nu < 0$.) So this yields

$$(5.34) \qquad p_h(m-i,h-i) = \binom{n+1}{h+1} - \Sigma_{j\geq 0}\binom{m-i+1}{h-i+j+1}\binom{n-m+i}{i-j}$$

We consider the case $i \geq h+1$. In the last sum, $j \geq 0$ as it stands. But if $j < 0$, then $h-i+j+1 < 0$ and hence $\binom{m-i+1}{h-i+j+1} = 0$. Thus the last sum may as well be extended to all integral values of j. It therefore equals

$$\Sigma_{\nu+\eta=h+1}\binom{m-i+1}{\nu}\binom{n-m+i}{\eta} = \binom{n+1}{h+1},$$

where the identity is obtained by comparing the coefficients of x^{h+1} in the expansions of $(1+x)^{n+1}$ and $(1+x)^{m-i+1} \cdot (1+x)^{n-m+i}$. Hence,

$$p_h(m-i,h-i) = 0 \quad \text{if} \quad i \geq h+1.$$

Therefore, extending the summation to all i in the last double sum of (5.33) means we must add to the left side of (5.33) the following quantity:

$$\Sigma_{m=k}^{n-1}\Sigma_{i=-\infty}^{-1} p_h(m-i,h-i)(-T_{m-i-1} + 2T_{m-i} - T_{m-i+1}).$$

But if $i < 0$, $p_h(m-i,h-i) = \binom{n+1}{h+1}$ because in (5.34), $\binom{n-m+i}{i-j} = 0$. So the above equals

$$\binom{n+1}{h+1}\Sigma_{m=k}^{n-1}\Sigma_{i=-\infty}^{-1}(-T_{m-i-1} + 2T_{m-i} - T_{m-i+1})$$

$$= \binom{n+1}{h+1}\Sigma_{m=k}^{n-1}(-T_m + T_{m+1})$$

$$= \binom{n+1}{h+1}(-T_k).$$

Hence (5.33) is equivalent to:

$$(5.35)\quad \sum_{A^h} m_k(A^h)$$
$$= \binom{n+1}{h+1}T_k - \Sigma_{m=k}^{n-1} \Sigma_{i=0}^{h} p_h(m-i,h-i)(S_{m-i} + E)$$
$$+ \Sigma_{m=k}^{n-1} \Sigma_{i=-\infty}^{+\infty} p_h(m-i,h-i)(-T_{m-i-1} + 2T_{m-i}$$
$$-T_{m-i+1}) + \mu(T^2).$$

We proceed to simplify the last sum. By choosing the subscript of the T's as the running subscript we may rewrite the double sum as

$$\Sigma_{m=k}^{n-1} \Sigma_{i=-\infty}^{+\infty} (-p_h(i+1,h-m+i+1) + 2p_h(i,h-m+i)$$
$$- p_h(i-1,h-m+i-1))T_i.$$

Now we apply the recursive relation among the binomial coefficients: $\binom{\mu}{\nu} + \binom{\mu}{\nu-1} = \binom{\mu+1}{\nu}$. This can be proved by inspecting the coefficient of x^ν in the expansions of $(1+x)^{\mu+1}$ and $(1+x)(1+x)^\mu$. This implies:

$$p_h(i,h-m+i) - p_h(i+1,h-m+i+1)$$
$$= \Sigma_{j\geq 0}\{\binom{i+2}{h-m+i+j+2}\binom{n-i-1}{m-i-j-1} - \binom{i+1}{h-m+i+j+1}\binom{n-i}{m-i-j}\}$$
$$= \Sigma_{j\geq 0}\{[\binom{i+1}{h-m+i+j+2} + \binom{i+1}{h-m+i+j+1}]\binom{n-i-1}{m-i-j-1}$$
$$- \binom{i+1}{h-m+i+j+1}[\binom{n-i-1}{m-i-j} + \binom{n-i-1}{m-i-j-1}]\}$$
$$= \Sigma_{j\geq 0}\{\binom{i+1}{h-m+i+j+2}\binom{n-i-1}{m-i-j-1} - \binom{i+1}{h-m+i+j+1}\binom{n-i-1}{m-i-j}\}$$
$$= - \binom{i+1}{h-m+i+1}\binom{n-i-1}{m-i}$$

In a similar fashion:

$$p_h(i-1,h-m+i-1) - p_h(i,h-m+i) = -\binom{i}{h-m+i}\binom{n-i}{m-i+1}$$

Hence the above double sum equals

$$\Sigma_{i=-\infty}^{+\infty} \Sigma_{m=k}^{n-1} \{\binom{i}{h-m+i}\binom{n-i}{m-i+1} - \binom{i+1}{h-m+i+1}\binom{n-i-1}{m-i}\}T_i$$

$$= \Sigma_{i=-\infty}^{+\infty} \Sigma_{m=k}^{n-1} \{\binom{i}{h-m+i}[\binom{n-i-1}{m-i+1} + \binom{n-i-1}{m-i}]$$

$$- [\binom{i}{h-m+i+1}\binom{i}{h-m+i}]\binom{n-i-1}{m-i}\}T_i$$

$$= \Sigma_{i=-\infty}^{+\infty} \Sigma_{m=k}^{n-1} \{\binom{i}{h-m+i}\binom{n-i-1}{m-i+1} - \binom{i}{h-m+i+1}\binom{n-i-1}{m-i}\}T_i$$

$$= \Sigma_{i=-\infty}^{+\infty} \{-\binom{i}{h-k+i+1}\binom{n-i-1}{k-i} + \binom{i}{h-n+1+i}\binom{n-i-1}{n-i}\}T_i$$

$$= \Sigma_{i=-\infty}^{+\infty} -\binom{i}{h-k+i+1}\binom{n-i-1}{k-i}T_i + \Sigma_{i=-\infty}^{+\infty} \binom{i}{h-n+i+1}\binom{n-i-1}{n-i}T_i$$

Now observe that

$$\binom{\mu}{\mu+1} = \begin{cases} 1 & \text{if} \quad \mu = -1 \\ 0 & \text{otherwise} . \end{cases}$$

So the second sum has every coefficient equal to zero except for $i = n$, but then $T_n = 0$. Thus the second sum vanishes identically. As to the first sum, remember that $h \leq k$ and $\binom{\mu}{\nu} = 0$ if $\nu < 0$. So $\binom{i}{h-k+i+1} = 0$ unless $i \geq (k-h)-1$ and $\binom{n-i-1}{k-i} = 0$ unless $k \geq i$. So the above equals

$$\Sigma_{i=k-h-1}^{k} -\binom{i}{h-k+i+1}\binom{n-i-1}{k-i}T_i .$$

We may now rewrite (5.35) in its final form:

$$(5.36) \quad \Sigma_{A^h} m_k(A^h) = \binom{n+1}{h+1}T_k - \Sigma_{m=k}^{n-1} \Sigma_{i=0}^{h} p_h(m-i,h-i)(S_{m-i} + E)$$
$$- \Sigma_{i=k-h-1}^{k}\binom{i}{h-k+i+1}\binom{n-i-1}{k-i}T_i + \mu(T^2).$$

When $h = k$, we claim that this reduces to

$$(5.37)\quad \sum_{A^k} m_k(A^k) = \binom{n+1}{k+1}T_k - \Sigma_{m=k}^{n-1} \Sigma_{i=0}^{k} p_k(m-i,k-i)(S_{m-i} + E) + \mu(T^2).$$

This is because

$$\Sigma_{i=-1}^{k} \binom{i}{i+1}\binom{n-i-1}{k-i}T_i$$

has every coefficient equal to zero due to the presence of $\binom{i}{i+1}$ except when $i = -1$. But then $T_{-1} = 0$. So the whole sum vanishes.

Making use of the First Corollary of Lemma 4.17, it is possible to derive a variant of (5.36) where only T_k appears on the right side. It goes as follows.

$$\begin{aligned} -\Sigma_{i=k-h-1}^{k} & \binom{i}{h-k+i+1}\binom{n-i-1}{k-i}T_i \\ &= \{-\Sigma_{i=k-h-1}^{k} \binom{i}{h-k+i+1}\binom{n-i-1}{k-i}\left(\frac{i+1}{k+1}\right)\}T_k \\ &\quad - \{\tfrac{1}{2} \Sigma_{i=k-h-1}^{k} \binom{i}{h-k+i+1}\binom{n-i-1}{k-i}(k-i)(i+1)\}E + \mu(T) \\ &= \{-\frac{k-h}{k+1} \Sigma_{i=k-h-1}^{k} \binom{i+1}{h-k+i+1}\binom{n-i-1}{k-i}\}T_k \\ &\quad - \{\frac{(n-k)(k-h)}{2} \Sigma_{i=k-h-1}^{k} \binom{i+1}{h-k+i+1}\binom{n-i-1}{k-i-1}\}E + \mu(T) \\ &= \{-\frac{k-h}{k+1} \Sigma_{i=k-h-1}^{k} \binom{i+1}{k-h}\binom{n-i-1}{n-k-1}\}T_k \\ &\quad - \{\frac{(n-k)(k-h)}{2} \Sigma_{i=k-h-1}^{k} \binom{i+1}{k-h}\binom{n-i-1}{n-k}\}E + \mu(T) \end{aligned}$$

In order to simplify the coefficients of T_k and E, we invoke the following identity: if p, q, r are positive integers, then

$$\Sigma_{\nu=q}^{q+r} \binom{p+q+r-\nu}{p}\binom{\nu}{q} = \binom{p+q+r+1}{r} .$$

To see this, recall that if m is a nonnegative integer, by definition, $\binom{-m}{p} = (-1)^p \frac{m\cdot(m+1)\cdot\cdots\cdot(m+p-1)}{p!}$. Hence for every nonnegative integer m,

$$\binom{-m}{p} = (-1)^p\binom{m+p-1}{m-1}$$

Now,

$$\begin{aligned}(1-x)^{-(p+1)} &= (1+(-x))^{-(p+1)} \\ &= \Sigma_\nu \binom{-(p+1)}{\nu}(-1)^\nu x^\nu \\ &= \Sigma_\nu \binom{p+\nu}{p} x^\nu .\end{aligned}$$

Similarly, $(1-x)^{-(q+1)} = \Sigma_\mu \binom{q+\mu}{q}x^\mu$ and $(1-x)^{-(p+q+2)} = \Sigma_\eta \binom{p+q+1+\eta}{p+q+1}x^\eta$. So comparing the coefficients of x^r in $(1-x)^{-(p+q+2)}$ and $(1-x)^{-(p+1)}\cdot(1-x)^{-(q+1)}$, we get the above identity.

Applying this identity, we obtain:

$$\begin{aligned}&- \Sigma_{i=k-h-1}^{k} \binom{i}{h-k+i+1}\binom{n-i-1}{k-i}T_i \\ &\quad = \{- \frac{(k-h)}{k+1} \Sigma_{j=k-h}^{k+1} \binom{n-j}{n-k-1}\binom{j}{k-h}\}T_k \\ &\qquad - \{\frac{(n-k)(k-h)}{2} \Sigma_{j=k-h}^{k+1} \binom{n-j}{n-k}\binom{j}{k-h}\}E + \mu(T) \\ &\quad = - \left(\frac{k-h}{k+1}\right)\binom{n+1}{h+1}T_k \\ &\qquad - \frac{(n-k)(k-h)}{2}\{\binom{n+1}{h} + \binom{k+1}{k-h}\binom{n-k-1}{n-k}\}E + \mu(T) \\ &\quad = - \frac{(k-h)}{k+1}\binom{n+1}{h+1}T_k - \frac{(n-k)(k-h)}{2}\binom{n+1}{h}E + \mu(T),\end{aligned}$$

where the last step is due to the fact that k never exceeds n-1, so that $\binom{n-k-1}{n-k} = 0$. Therefore from (5.36) we deduce

$$(5.38)\quad \sum_{A^h} m_k(A^h) = \frac{(h+1)}{(k+1)}\binom{n+1}{h+1}T_k - \Sigma_{m=k}^{n-1}\Sigma_{i=0}^{h} p_h(m-i,h-i)S_{m-i}$$
$$- \{\frac{(n-k)(k-h)}{2}\binom{n+1}{h} + \Sigma_{m=k}^{n-1}\Sigma_{i=0}^{h} p_h(m-i,h-i)\}E + \mu(T^2)$$

where use has been made of Lemma 4.16(i) and (v) to get $\mu(T) + \mu(T^2) = \mu(T^2)$. (5.36)-(5.38) <u>constitute the defect relations for</u> $0 \leq h \leq k$.

To complete the picture, let us deduce also the defect relations for the case $k \leq h \leq n-1$. Because the details are somewhat similar to the preceding, we will be brief. We first need the counterpart of Lemma 5.10.

<u>Lemma 5.11</u>. Let B^k be a fixed k-dimensional projective subspace of $P_n\mathbb{C}$. For a system of h-dimensional projective subspaces $\{A^h\}$ in general position, the number of A^h with the property that every $A^\ell \supseteq A^h$ (ℓ a fixed integer and $k \leq \ell$, $h \leq \ell$) satisfies $B^k \rfloor A^\ell = 0$, does not exceed

$$p_{n-h-1}(n-k-1,n-\ell-1) = \binom{n+1}{n-h} - \Sigma_{j\geq 0}\binom{n-k}{n-\ell+j}\binom{k+1}{\ell-h-j} .$$

This lemma reduces to Lemma 5.10 if we replace each A^h by its polar space $A^\perp$ ($\dim A^\perp = n-h-1$) and B^k by its polar space $B^\perp$ ($\dim B^\perp = n-k-1$).

We now suppose that we are given $A^h = e_0 \wedge \cdots \wedge e_h$, where $e_0,\ldots,e_h$ are part of an O.N. basis $e_0,\ldots,e_n$ in $\mathbb{C}^{n+1}$. Let $A^{\ell+1} \supseteq A^\ell \supseteq A^h$ be projective subspaces corresponding to the subspaces of $\mathbb{C}^{n+1}$ spanned by an arbitrary choice of $(\ell+2)$ and $(\ell+1)$ members of $\{e_0,\ldots,e_n\}$. We apply Theorem 5.7 with $k \leq \ell \leq n-1$:

$$(1-\alpha)\int_{r_0}^{r} dt \int_{r_0}^{t} ds \int_{\partial V[s]} \frac{|X^{k-1} \rfloor A^{\ell}|^2 |X^{k+1} \rfloor A^{\ell+1}|^2}{|X^k \rfloor A^{\ell+1}|^4} \cdot \left(\frac{|X^k \rfloor A^{\ell+1}|}{|X^k \rfloor A^{\ell}|}\right)^{2\alpha} * d\tau$$

$$\leq CT_k(r) + C'$$

where $0 < \alpha < 1$, and C, C' are positive constants independent of α, A^{ℓ} and $A^{\ell+1}$. Proceeding in similar manner as above, we first obtain:

$$(1-\alpha)\int_{r_0}^{r} dt \int_{r_0}^{t} ds \int_{\partial V[s]} \frac{|X^{k-1}|^2 |X^{k+1}|^2}{|X^k|^4} \cdot \frac{\mathrm{Min}\left(\frac{|X^k|^{2\alpha}}{|X^k \rfloor A^{\ell}|^{2\alpha}} : \frac{|X^{k-1}|^2}{|X^{k-1} \rfloor A^{\ell}|^2}\right)}{\mathrm{Min}\left(\frac{|X^{k+1}|^2}{|X^{k+1} \rfloor A^{\ell+1}|^2} : \frac{|X^k|^2}{|X^k \rfloor A^{\ell+1}|^2}\right)} * d\tau$$

$$\leq CT_k(r) + C'$$

where C, C' are new positive constants still independent of α, and the minimum refers to the finite number of A^{ℓ} and $A^{\ell+1}$ containing A^h. Letting A^h range over a finite system in general position, we obtain the analogue of (5.32):

$$\sum_{A^h} \frac{1}{2\pi} \int_{\partial V[s]} \log \mathrm{Min}\left(\frac{|X^k|}{|X^k \rfloor A^{\ell}|} : \frac{|X^{k-1}|}{|X^{k-1} \rfloor A^{\ell}|}\right) * d\tau$$

$$- \sum_{A^h} \frac{1}{2\pi} \int_{\partial V[s]} \log \mathrm{Min}\left(\frac{|X^{k+1}|}{|X^{k+1} \rfloor A^{\ell+1}|} : \frac{|X^k|}{|X^k \rfloor A^{\ell+1}|}\right) * d\tau$$

$$= p_{n-h-1}(n-k-1, n-\ell-1)(-E-S_k-T_{k-1}+2T_k-T_{k+1}) + \mu(T^2) .$$

Let $\Psi(k,\ell) = \sum_{A^h} \frac{1}{2\pi} \int_{\partial V[s]} \log \operatorname{Min}\left(\frac{|X^k|}{|X^k \lrcorner A^\ell|} : \frac{|X^{k-1}|}{|X^{k-1} \lrcorner A^\ell|}\right) * d\tau,$

then the above may be written

$$(5.39)^k_\ell \quad \Psi(k,\ell) - \Psi(k+1,\ell+1)$$
$$= p_{n-h-1}(n-k-1,n-\ell-1)(-E-S_k-T_{k-1}+2T_k-T_{k+1}) + \mu(T^2).$$

Adding $(5.39)^k_\ell, (5.39)^{k+1}_{\ell+1}, \ldots, (5.39)^{k+n-\ell+1}_{n-1}$, and noting that $\Psi(k+n-\ell,n) = 0$, we get

$$\sum_{A^h} \frac{1}{2\pi} \int_{\partial V[s]} \log \operatorname{Min}\left(\frac{|X^k|}{|X^k \lrcorner A^\ell|} : \frac{|X^{k-1}|}{|X^{k-1} \lrcorner A^\ell|}\right) * d\tau$$
$$= \Sigma_{i=0}^{n-\ell-1}\, p_{n-h-1}(n-k-i-1,n-h-i-1)(-E-S_{k+i}-T_{k+i+1}$$
$$+ 2T_{k+i}-T_{k+i-1}) + \mu(T^2)$$

If $\ell = h$, no minimum need be taken, so this reduces to

$$\sum_{A^h} \{m_k(A^h) - m_{k-1}(A^h)\}$$
$$= \Sigma_{i=0}^{n-h-1}\, p_{n-h-1}(n-k-i-1,n-h-i-1)(-E-S_{k+i}-T_{k+i-1}$$
$$+ 2T_{k+i}-T_{k+i+1}) + \mu(T^2),$$

where we have written $m_k(A^h)$ in place of $m_k(r,A^h)$ as usual. Summing over k from 0 to k and noting that $m_{-1}(A^h) \equiv 0$, we have:

$$\sum_{A^h} m_k(A^h) = -\Sigma_{m=0}^{k}\, \Sigma_{i=0}^{n-h-1}\, p_{n-h-1}(n-m-i-1,n-h-i-1)(E+S_{m+i})$$
$$+ \Sigma_{m=0}^{k}\, \Sigma_{i=0}^{n-h-1}\, p_{n-h-1}(n-m-i-1,n-h-i-1)$$
$$\cdot (-T_{m+i-1}+2T_{m+i}-T_{m+i+1}) + \mu(T^2)$$

This is valid when $k \leq h$. We now extend the last double sum over all i, setting as before $T_i = 0$ when $i < 0$ and when $i > n-1$. The coefficient vanishes when $i \geq n-h$ and equals $\binom{n+1}{n-h} = \binom{n+1}{h+1}$ when $i < 0$. So after we have extended the sum to all i, we should add to the left the following quantity:

$$\Sigma_{m=0}^{k} \Sigma_{i=-\infty}^{-1} p_{n-h-1}(n-m-i-1, n-h-i-1)(-T_{m+i-1}+2T_{m+i}-T_{m+i+1})$$

$$= \binom{n+1}{h+1} \Sigma_{m=0}^{k} \Sigma_{i=-\infty}^{-1} (-T_{m+i-1}+2T_{m+i}-T_{m+i+1})$$

$$= - \binom{n+1}{h+1} T_k.$$

We have therefore obtained the following:

$$(5.40) \quad \Sigma_{A^h} m_k(A^h) = \binom{n+1}{h+1} T_k - \Sigma_{m=0}^{k} \Sigma_{i=0}^{n-h-1} p_{n-h-1}(n-m-i-1, n-h-i-1) \cdot (E+S_{m+i}) + \Sigma_{m=0}^{k} \Sigma_{i=-\infty}^{+\infty} p_{n-h-1}(n-m-i-1, n-h-i-1) \cdot (-T_{m+i-1}+2T_{m+i}-T_{m+i+1}) + \mu(T^2)$$

It remains to simplify the last double sum. Let $\alpha = n-h-1$ and let $\beta = n-1-m$. Then this double sum may be written as:

$$(5.41) \quad \Sigma_{\beta=n-1-k}^{n-1} \Sigma_{i=-\infty}^{+\infty} p_\alpha(\beta-i, \alpha-i)(-T_{n-1-(\beta-i-1)}+2T_{n-1-(\beta-i)} -T_{n-1-(\beta-i+1)})$$

It may be recalled that previously we have computed the last infinite double sum of (5.35) and found that

$$(5.42) \quad \Sigma_{m=k}^{n-1} \Sigma_{i=-\infty}^{+\infty} p_h(m-i, h-i)(-T_{m-i-1}+2T_{m-i}-T_{m-i+1}) = \Sigma_{i=k-h-1}^{k} -\binom{i}{h-k+i+1}\binom{n-i-1}{k-i} T_i.$$

The left side of (5.42) would be identical with (5.41) if only we replace h by α, k by $(n-1-k)$ and T_j by T_{n-1-j}. So (5.41) equals

$$\Sigma_{i=n-k-1-\alpha-1}^{n-k-1} -\binom{i}{\alpha-(n-k-1)+i+1}\binom{n-i-1}{(n-k-1)-i}T_{n-1-i}$$

$$= \Sigma_{i=h-k-1}^{n-k-1} -\binom{i}{k-h+i+1}\binom{n-i-1}{n-i-k-1}T_{n-1-i}$$

$$= \Sigma_{j=k}^{n-h+k} -\binom{n-1-j}{k-h+n-j}\binom{j}{j-k}T_j$$

$$= \Sigma_{j=k}^{n-h-k} -\binom{n-1-j}{h-k-1}\binom{j}{k}T_j$$

Substituting this into (5.40), we obtain the counterpart of (5.36):

$$(5.43) \quad \sum_{A^h} m_k(A^h) = \binom{n+1}{h+1}T_k - \Sigma_{m=0}^{k} \Sigma_{i=0}^{n-k-1} p_{n-h-1}(n-m-i-1,n-h-i-1) \cdot (E+S_{m+i}) - \Sigma_{i=k}^{n-h+k} \binom{n-1-i}{h-k-1}\binom{i}{k}T_i + \mu(T^2) .$$

This is valid for $k \leq h$. When $k = h$, (5.43) again reduces to (5.37).

Again we can transform the last sum using the First Corollary of Lemma 4.17 so that only T_k appears.

$$- \Sigma_{i=k}^{n-h+k} \binom{n-1-i}{h-k-1}\binom{i}{k}T_i$$

$$= \{-\Sigma_{i=k}^{n-h+k} \left(\frac{n-i}{n-k}\right)\binom{n-1-i}{h-k-1}\binom{i}{k}\}T_k$$

$$- \{\frac{1}{2} \Sigma_{i=k}^{n-h+k} \binom{n-1-i}{h-k-1}\binom{i}{k}(i-k)(n-i)\}E + \mu(T)$$

$$= \{- \frac{h-k}{n-k} \Sigma_{i=k}^{n-h+k} \binom{n-i}{h-k}\binom{i}{k}\}T_k$$

$$- \{\frac{(k+1)(h-k)}{2} \Sigma_{i=k}^{n-h+k} \binom{n-i}{h-k}\binom{i}{k+1}\}E + \mu(T)$$

$$= -\frac{(h-k)}{(n-k)}\binom{n+1}{h+1}T_k - \{\tfrac{1}{2}(k+1)(h-k)\binom{n+1}{h+2}\}E + \mu(T)$$

where we have used the previously proved identity:

$$\Sigma_{\nu=q}^{q+r}\binom{p+q+r-\nu}{p}\binom{\nu}{q} = \binom{p+q+r+1}{r}.$$

Substituting into (5.43) the above, we have arrived at the counterpart of (5.38):

$$(5.44)\quad \Sigma_{A^h} m_k(A^h) = \frac{(n-h)}{(n-k)}\binom{n+1}{h+1}T_k$$
$$- \Sigma_{m=0}^{k}\,\Sigma_{i=0}^{n-k-1}\,p_{n-h-1}(n-m-i-1,n-h-i-1)S_{m+i}$$
$$- \{\frac{(k+1)(h-k)}{2}\binom{n+1}{h+2}$$
$$+ \Sigma_{m=0}^{k}\,\Sigma_{i=0}^{n-k-1}\,p_{n-h-1}(n-m-i-1,n-h-i-1)\}E$$
$$+ \mu(T^2)$$

(5.43) and (5.44) are the defect relations for the case $(n-1) \geq h \geq k$. We now summarize the above into a comprehensive theorem, which is the main result of the whole development.

<u>Theorem 5.12 (Defect Relations)</u>. Let $x: V \to P_n\mathbb{C}$ be a nondegenerate holomorphic curve and let V admit a harmonic exhaustion. Let further $\{A^h\}$ be a finite system of h-dimensional projective subspaces of $P_n\mathbb{C}$ in general position. Then the generalized compensating terms $m_k(A^h) \equiv m_k(r,A^h)$ of (4.29) satisfy the following inequalities: If $0 \leq h \leq k$, then

$$\Sigma_{A^h} m_k(A^h) = \binom{n+1}{h+1}T_k - \Sigma_{m=k}^{n-1}\,\Sigma_{i=0}^{h}\,p_h(m-i,h-i)(S_{m-i}+E)$$
$$- \Sigma_{i=k-h-1}^{k}\binom{i}{h-k+i+1}\binom{n-i-1}{k-i}T_i + \mu(T^2)$$

$$= \binom{h+1}{k+1}\binom{n+1}{h+1}T_k - \Sigma_{m=k}^{n-1} \Sigma_{i=0}^{h} p_h(m-i,h-i)S_{m-i}$$

$$- \{\frac{(n-k)(k-h)}{2}\binom{n+1}{h} + \Sigma_{m=k}^{n-1} \Sigma_{i=0}^{h} p_h(m-i,h-i)\}E + \mu(T^2)$$

If $k \leq h \leq n-1$, then

$$\sum_{A^h} m_k(A^h) = \binom{n+1}{h+1}T_k - \Sigma_{m=0}^{k} \Sigma_{i=0}^{n-k-1} p_{n-h-1}(n-m-i-1,n-h-i-1) \cdot (E+S_{m+i})$$

$$- \Sigma_{i=k}^{n-h+k} \binom{n-1-i}{h-k-1}\binom{i}{k}T_i + \mu(T^2)$$

$$= \left(\frac{n-h}{n-k}\right)\binom{n+1}{h+1}T_k$$

$$- \Sigma_{m=0}^{k} \Sigma_{i=0}^{n-k-1} p_{n-h-1}(n-m-i-1,n-h-i-1)S_{m+i}$$

$$- \{\frac{1}{2}(k+1)(h-k)\binom{n+1}{h+2} + \Sigma_{m=0}^{k} \Sigma_{i=0}^{n-k-1} p_{n-h-1} \cdot (n-m-i-1,n-h-i-1)\}E + \mu(T^2).$$

Finally, if $h = k$, then

$$\sum_{A^k} m_k(A^k) = \binom{n+1}{k+1}T_k - \Sigma_{m=k}^{n-1} \Sigma_{i=0}^{k} p_k(m-i,k-i)(S_{m-i}+E) + \mu(T^2).$$

As we mentioned before, the last conclusion of the preceding theorem is the most important in application. We can rephrase it in an essentially equivalent way, as follows. Let V admit an infinite harmonic exhaustion (Definition 2.4). For each k-dimensional projective subspace A^k of $P_n\mathbb{C}$, we define the <u>defect of</u> A^k to be:

$$\delta_k(A^k) = \liminf_{r\to\infty} \left(1 - \frac{N_k(r,A^k)}{T_k(r)}\right)$$

Then δ_k is a measurable real-valued function on $G(n,k)$.

Clearly $\delta_k \leq 1$. From (4.25) $(N_k(r,A^k) < T_k(r) + c_k)$, it follows that $0 \leq \delta_k$. (See §5 of Chapter II). Thus δ_k: $G(n,k) \to [0,1]$. Let $A^\perp$ be the polar space of A^k. From the definition of $N_k(r,A^k)$ (see the paragraph preceding Theorem 4.8 and the discussion after (4.4)), if $_kx(p) \cap A^\perp = \emptyset$ for all $p \in V$, then $N_k(r,A^k) = 0$ for all r and so $\delta_k(A^k) = 1$. $\delta_k(A^k) = 0$ is then to be interpreted as $_kx(p) \cap A^\perp \neq \emptyset$ for "many" $p \in V$. We now show that in many cases, $\delta_k \equiv 0$ except on a countable subset of $G(n,k)$.

Theorem 5.13 (Defect Relations). Let $x\colon V \to P_n\mathbb{C}$ be a nondegenerate holomorphic curve and suppose either (i) $V = \mathbb{C}$ or $\mathbb{C} - \{0\}$, or else (ii) $V =$ compact $M - \{a_1,\ldots,a_m\}$ $(a_i \in M)$ and x is transcendental. If $\{A^k\}$ is a finite system of k-spaces in general position $(0 \leq k \leq n-1)$, then

$$\sum_{A^k} \delta_k(A^k) \leq \binom{n+1}{k+1}$$

Proof. Suppose not, then

$$\liminf_{r\to\infty} \sum_{A^k} \left(1 - \frac{N_k(r,A^k)}{T_k(r)}\right) \geq \binom{n+1}{k+1} + \epsilon ,$$

where $\epsilon > 0$. Thus outside a compact set,

$$\sum_{A^k} \left(1 - \frac{N_k(r,A^k)}{T_k(r)}\right) \geq \binom{n+1}{k+1} + \epsilon ,$$

or $\sum_{A^k} \{T_k(r) - N_k(r,A^k)\} \geq \{\binom{n+1}{k+1} + \epsilon\}T_k(r)$, or in view of (4.16), $\sum_{A^k} m_k(r,A^k) \geq \{\binom{n+1}{k+1} + \epsilon\}T_k(r)$. By Lemma 4.16(i) and the last conclusion of the preceding theorem:

$$\{\binom{n+1}{k+1} + \epsilon\}T_k(r) = \binom{n+1}{k+1}T_k - \Sigma_{k,i}\ p_k(m-i,k-i)(S_{m-i}+E) + \mu(T^2)$$

$$\leftrightarrow \epsilon T_k + \Sigma_{k,i}\ p_k(m-i,k-i)(S_{m-i}+E) = \mu(T^2).$$

Now $S_{m-i} \geq 0$ by its definition, so Lemma 4.16(iv) implies

$$\epsilon T_k + (\Sigma_{k,i}\ p_k(m-i,k-i))E = \mu(T^2).$$

Since $\epsilon > 0$,

$$T_k + \frac{1}{\epsilon}(\Sigma_{k,i}\ p_k(m-i,k-i))E = \mu(T^2),$$

by Lemma 4.16(iii). Note that each $p_k(m-i,k-i)$ is a positive constant, so the coefficient of E is a positive constant. This contradicts the last conclusion of Theorem 4.24. Q.E.D.

We mention in passing that if we only know that V has an infinite harmonic exhaustion, we can still obtain defect relations. In fact, define

$$\chi_k = (\Sigma_{m=k}^{n-1} \Sigma_{i=0}^{k}\ p_k(m-i,k-i))\limsup_{r\to\infty} \frac{-E(r)}{T_k(r)}$$

Then one can prove that for a finite system of k-spaces $\{A^k\}$ in general position, the following holds:

(5.45) $$\sum_{A^k} \delta_k(A^k) \leq \binom{n+1}{k+1} + \chi_k,$$

and if one of $\{\chi_0,\ldots,\chi_{n-1}\}$ is finite, so are the others. Because this seems to be too complicated to be of much use, we leave it as an exercise to the reader.

§5. Ahlfors mentioned at the beginning of his paper [1] that the equidistribution theory of holomorphic curves suffers from a lack of applications. The situation has not much improved in this respect in the intervening thirty years. In this concluding section of the notes: we attempt to give a few simple consequences of the defect relations, some of which are classical.

We have as usual a nondegenerate holomorphic curve $x: V \to P_n\mathbb{C}$. For each $p \in V$, ${}_kx(p)$ is of course a k-dimensional projective subspace of $P_n\mathbb{C}$, and we call it a <u>k-dimensional osculating space</u> of x at p.

<u>Proposition 5.14</u>. Let $x: V \to P_n\mathbb{C}$ be a nondegenerate holomorphic curve. If either (i) $V = \mathbb{C}$ or $\mathbb{C} - \{0\}$, or (ii) x is transcendental and V is a compact Riemann surface with a finite number of points deleted, then given $\{\binom{n+1}{k+1} + 1\}$ k-dimensional projective subspaces of $P_n\mathbb{C}$ in general position, $(0 \leq k \leq n-1)$, at least one of them meets an $(n-k-1)$-dimensional osculating spaces of x. In particular, given $(n+2)$ hyperplanes in general position, $x(V)$ must intersect one of them.

<u>Proof</u>. Let $\{A^k\}$ be the original finite system of k-spaces and let $\{A^\perp\}$ be the corresponding system of $(n-k-1)$-spaces formed from the polar spaces of $\{A^k\}$. Each $A^\perp$ has the same Grassmannian coordinates as the corresponding A^k, so if $\{A^k\}$ are in general position, Lemma 5.8(b) shows that the $\{A^\perp\}$ are also in general position. If ${}_{(n-k-1)}x(p)$ does

not meet $\{A^k\}$ for any $p \in V$, then $N_{n-k-1}(r, A^\perp) = 0$ for each $A^\perp$ and each r, and therefore $\delta_{n-k-1}(A^\perp) = 1$ for each $A^\perp$. Since there are $\{\binom{n+1}{k+1} + 1\}$ $(= \{\binom{n+1}{n-k} + 1\})$ such $A^\perp$'s, we obtain $\sum_{A^\perp} \delta_{n-k-1}(A^\perp) = \binom{n+1}{n-k} + 1$. This contradicts Theorem 5.13. Q.E.D.

From this proposition follows the classical theorem of Borel. We formulate it in this fashion. Given holomorphic function $x_0, \ldots, x_{n+1}$ on V, we call a <u>linear relation</u> among them an identity of the form:

$$a_0 x_0 + \cdots + a_{n+1} x_{n+1} = 0$$

where $a_0, \ldots, a_{n+1}$ are complex numbers with <u>at least</u> one a_i not equal to zero. We call it a <u>special linear relation</u> if and only if <u>none</u> of the $a_0, \ldots, a_{n+1}$ is equal to zero.

<u>Proposition 5.15 (E. Borel)</u>. Let $V = \mathbb{C}$ or $\mathbb{C} - \{0\}$ and let $(n+2)$ nowhere zero holomorphic functions $x_0, \ldots, x_{n+1}$ be given on V, $(n \geq 1)$. Then a special linear relation among $x_0, \ldots, x_{n+1}$ implies a linear relation among any $(n+1)$ of them.

<u>Proof</u>. Let the special linear relation be $a_0 x_0 + \cdots + a_{n+1} x_{n+1} = 0$. After renumbering if necessary, let us say that we want a linear relation among $x_0, \ldots, x_n$. Consider the holomorphic mapping $\tilde{x}: V \to \mathbb{C}^{n+1} - \{0\}$ where $\tilde{x} = (x_0, \ldots, x_n)$. If $\pi: \mathbb{C}^{n+1} - \{0\} \to P_n\mathbb{C}$ as usual, then $x \equiv \pi \circ \tilde{x}$ is of course a holomorphic curve in $P_n\mathbb{C}$. Obviously, x never meets the $(n+2)$ hyperplanes in general position given by: $z_0 = 0, z_1 = 0, \ldots, z_r$

and $a_0 z_0 + \cdots + a_n z_n = 0$. Proposition 5.14 implies that x must be degenerate and so the vector-valued function $\tilde{x}$ takes value in an n-dimensional subspace of $\mathbb{C}^{n+1}$ defined by, say, $b_0 z_0 + \cdots + b_n z_n = 0$. Hence $b_0 x_0 + \cdots + b_n x_n = 0$

Q.E.D.

Proposition 5.16. Let V be a compact Riemann surface M with a finite number of points deleted and let $(n+2)$ nowhere zero holomorphic functions $x_0, \ldots, x_{n+1}$ be given on V $(n \geq 1)$. Then a special linear relation among $x_0, \ldots, x_{n+1}$ implies either that a linear relation exists among $(n+1)$ of them, or that every quotient x_i/x_j $(0 \leq i, j \leq n+1)$ can be extended to a meromorphic function on M, or both.

Proof. Of course we may exclude on the outset the cases $V = \mathbb{C}$ or $\mathbb{C} - \{0\}$. Let $V = M - \{a_1, \ldots, a_m\}$, where M is a compact Riemann surface and $a_i \in M$. Our first observation is that if the holomorphic curve $x\colon V \to P_n\mathbb{C}$ induced by $\tilde{x}\colon V \to \mathbb{C}^{n+1}$ (where $\tilde{x} = (x_0, \ldots, x_n)$) is not transcendental, then every quotient x_α/x_β $(0 \leq \alpha, \beta \leq n)$ is extendable to a meromorphic function on M.

To prove this, note that actually $\tilde{x}\colon V \to \mathbb{C}^{n+1} - \{0\}$, so that $x = \pi \circ \tilde{x}$, where $\pi\colon \mathbb{C}^{n+1} - \{0\} \to P_n\mathbb{C}$. Let us show for instance that $x_1/x_0, \ldots, x_n/x_0$ are all extendable over a_1 to be meromorphic functions. So take a neighborhood U of a_1 and take the usual coordinate neighborhood $U_0 = \{[z_0, \ldots, z_n]\colon z_0 \neq 0\}$ on $P_n\mathbb{C}$ with coordinate function $\zeta\colon U_0 \to \mathbb{C}^n$ such that $\zeta([z_0, \ldots, z_n]) = (z_1/z_0, \ldots, z_n/z_0)$. Because x_0 never vanishes, $x(U - \{a_1\}) \subseteq U_0$, so that $\zeta \circ x$ is well-defined on $U - \{a_1\}$. Quite obviously, $\zeta \circ x\colon U - \{a_1\} \to \mathbb{C}^n$ is such that $\zeta \circ x = (\frac{x_1}{x_0}, \ldots, \frac{x_n}{x_0})$. Since by assumption x is not transcendental, Lemma 4.23 implies that x is extendable to a holomorphic mapping defined on all of U into $P_n\mathbb{C}$. (The proof of that part of the lemma remains valid regardless of whether x is nondegenerate or not). By the proof of Lemma 3.3, $\zeta \circ x$ is then extendable to n meromorphic functions on U, which proves our claim.

Now to the proof of our lemma. Let the special linear relation be $a_0 x_0 + \cdots + a_{n+1} x_{n+1} = 0$. It suffices to prove that if there is no linear relation among any $(n+1)$ of $x_0, \ldots, x_{n+1}$, then every quotient x_i/x_j can be extended to a meromorphic function on all of M $(0 \leq i, j \leq n+1)$. Let us show that $\frac{x_1}{x_0}, \ldots, \frac{x_n}{x_0}$ can be so extended, the proof of the others being similar. Consider therefore the curve $x\colon V \to P_n\mathbb{C}$ induced by $\tilde{x}\colon V \to \mathbb{C}^{n+1} - \{0\}$, where $\tilde{x} = (x_0, \ldots, x_n)$. Since actually $\pi \circ \tilde{x} = x$, it is easy to see that x does not meet any of the $(n+2)$ hyperplanes given by: $z_0=0, \ldots, z_n=0$, and $a_0 z_0 + \cdots + a_n z_n = 0$. These hyperplanes are in general position, and so Proposition 5.14 says that either x is degenerate or x is not transcendental, or both. Since there is no linear relation among $x_0, \ldots, x_n$, $\tilde{x}$ does not take value in a hyperplane of $\mathbb{C}^{n+1}$; we have to conclude then that x is not transcendental. By our initial observation, $\frac{x_1}{x_0}, \ldots, \frac{x_n}{x_0}$, are all extendable to all of M as meromorphic functions. Q.E.D.

One can find an application of this theorem to the uniqueness problem of meromorphic functions in the thesis of Edwardine Schmid (Berkeley 1969). Our next proposition depends on a lemma which has not been completely proved.

Conjectural Lemma 5.17. Let $H_1, \ldots, H_{n+2}$ be $(n+2)$ hyperplanes in general position in $P_n\mathbb{C}$, $n \geq 2$. Consider the set of k-dimensional subspaces A^k $(A^k \not\subseteq H_i,\ i = 1, \ldots, n+2)$ of $P_n\mathbb{C}$ which have the property: $\{A^k \cap H_i : i = 1, \ldots, n+2\}$ contains fewer than $(k+2)$ hyperplanes of A^k in general position. (A hyperplane of A^k is of course a $(k-1)$-space in $P_n\mathbb{C}$). Then the union of all such A^k's as k runs through $1, \ldots, n-1$, is the union of a finite number of distinct proper projective subspaces of $P_n\mathbb{C}$.

The proof of this lemma for $n = 2,3,4$ is not difficult; it is also relatively easy to prove that the number of hyperplanes H such that $\{H \cap H_i : i = 1, \ldots, n+2\}$ contains fewer than $(n-1)+2$ hyperplanes of H in general position is finite in number. However, the general proof for $n = 5$ starts to get very long and I have not carried it through. In any case, the following proposition follows from this conjecture.

Proposition 5.18. Suppose $x: \mathbb{C}^m \to P_n\mathbb{C}$ is a holomorphic mapping such that $x(\mathbb{C}^m)$ avoids $(n+2)$ hyperplanes in general position, then $x(\mathbb{C}^m)$ lies in a proper projective subspace of $P_n\mathbb{C}$.

Proof. Let us say that $x(\mathbb{C}^m)$ avoids $H_1, \ldots, H_{n+2}$ and the latter are in general position. We may assume on the outset that x is not a constant map. Let p be any point of $\mathbb{C}^m$. Consider the set of complex lines passing through p; their union is all of $\mathbb{C}^m$. Hence the restriction of x to at least one of them should be nonconstant. So there is no harm in assuming that $x: \mathbb{C}_p \to P_n\mathbb{C}$ is nonconstant, where we have used the notation: if $p = (p_1,\ldots,p_m)$, $\mathbb{C}_p = \{(z,p_2,\ldots,p_m): z \in \mathbb{C}\}$. In other words, $\mathbb{C}_p$ is the complex line through p which is parallel to $(1,0,\ldots,0)$. Now $x: \mathbb{C}_p \to P_n\mathbb{C}$ is a nonconstant holomorphic curve that avoids $(n+2)$ hyperplanes of $P_n\mathbb{C}$ in general position and is therefore degenerate, by Proposition 5.14. Say it lies in an ℓ-space A^ℓ of $P_n\mathbb{C}$, $1 \leq \ell \leq n-1$. If we choose ℓ to be the smallest integer for which $x(\mathbb{C}_p) \subseteq A^\ell$, then $x: \mathbb{C}_p \to A^\ell$ is a nondegenerate holomorphic curve. Since $x(\mathbb{C}_p)$ still avoids $\{A^\ell \cap H_i: i = 1,\ldots,n+2\}$, the latter must contain fewer than $(\ell+2)$ hyperplanes of A^ℓ in general position by Proposition 5.14. We are thus forced to conclude that A^ℓ lies in the union of the finite number of proper projective subspaces of $P_n\mathbb{C}$ guaranteed by Conjectural Lemma 5.17, say, $A_1,\ldots,A_s$. Let us vary p over an open set U' of $\mathbb{C}^m$. If U' is sufficiently small, we can clearly assume that $x: \mathbb{C}_p \to P_n\mathbb{C}$ is nonconstant for each $p \in U'$. The union of all $\mathbb{C}_p$, $p \in U'$, is an open set U of $\mathbb{C}^m$, and the preceding argument obviously implies that $x(U) \subseteq A_1 \cup \cdots \cup A_s$. We now use induction on s to show that $x(U)$ actually lies in one of the $A_1,\ldots,A_s$. If $s = 1$, there is nothing to prove. So suppose $x(U) \subseteq A_1 \cup \cdots \cup A_{s-1}$ implies that $x(U)$ lies in one of the $A_1,\ldots,A_{s-1}$.

We now show that if $x(U) \subseteq A_1 \cup \cdots \cup A_s$, then $x(U) \subseteq A_j$, for some j between 1 and s. Suppose for some $q \in U$, $x(q) \in A_s - \{A_1 \cup \cdots \cup A_{s-1}\}$, then by continuity, x carries a sufficiently small neighborhood of q into A_s that is disjoint from $A_1 \cup \cdots \cup A_{s-1}$. But if x carries an open set into A_s, by the holomorphy of x, $x(\mathbb{C}^m) \subseteq A_s$, and we are done in this case. So we may assume that $x(q) \notin A_s - \{A_1 \cup \cdots \cup A_{s-1}\}$ for any $q \in U$. Then $x(U) \subseteq A_1 \cup \cdots \cup A_{s-1}$ and our induction hypothesis implies that $x(U) \subseteq A_i$, $1 \leq i \leq s-1$. The holomorphy of x now implies that $x(\mathbb{C}^m) \subseteq A_i$. Q.E.D.

<u>Corollary</u>. Let $H_1, \ldots, H_{n+2}$ be $(n+2)$ hyperplanes of $P_n\mathbb{C}$ in general position and $x\colon \mathbb{C}^n \to P_n\mathbb{C}$ is a holomorphic mapping whose differential is nonsingular somewhere. Then $x(\mathbb{C}^n)$ must intersect one of $H_1, \ldots, H_{n+2}$.

This generalizes the classical Picard theorem.

We wish to generalize the Picard theorem in yet another way. For each positive integer n, let $\rho(n)$ be the positive integer greater than or equal to $(2n+1)$ such that the following lemma is valid:

<u>Lemma 5.19</u>. Let $H_1, \ldots, H_{\rho(n)}$ be $\rho(n)$ hyperplanes of $P_n\mathbb{C}$ in general position and let A^k $(1 \leq k \leq n-1)$ be a k-dimensional projective subspace of $P_n\mathbb{C}$ not contained in any H_i. Then $\{A^k \cap H_i\colon i = 1, \ldots, \rho(n)\}$ contains at least $(k+2)$ hyperplanes of A^k in general position.

For $n \leq 4$, it can be shown by a straightforward case by case examination that $\rho(n) = 2n+1$ would do. The determination of the smallest possible $\rho(n)$ for a general n remains unsettled. In any case, on the basis of this lemma, we can prove the following.

Proposition 5.20. Let $x: \mathbb{C}^m \to P_n\mathbb{C}$ be a holomorphic mapping such that $x(\mathbb{C}^m)$ fails to intersect $\rho(n)$ hyperplanes of $P_n\mathbb{C}$ in general position. Then x must reduce to a constant.

When $m = n = 1$, this says that a meromorphic function on the Gaussian plane must be a constant if it omits three distinct points of the Riemann sphere (because we have remarked above that $\rho(1) = 3$ would do), which is Picard's theorem again.

Proof. It suffices to prove the proposition for a holomorphic $x: \mathbb{C} \to P_n\mathbb{C}$. Since $\rho(n) \geq (n+2)$, Proposition 5.14 says that x must be degenerate. If x is not a constant, then $x(\mathbb{C})$ lies in a projective subspace A^k, $1 \leq k \leq n-1$. Choose A^k so that it is the smallest subspace containing $x(\mathbb{C})$, then $x: \mathbb{C} \to A^k$ is a nondegenerate holomorphic curve which avoids all the hyperplanes $\{A^k \cap H_i: i = 1,\ldots,\rho(n)\}$. But at least $(k+2)$ of the latter are in general position by Lemma 5.19, so this contradicts Proposition 5.14. Q.E.D.

Proposition 5.20 is interesting in view of Kobayashi's theory of hyperbolic (complex) manifolds. (cf. a forthcoming monograph of his; see also J. Math. Soc. Japan 1967, 460-480).

A hyperbolic manifold has most of the essential properties of a bounded domain, including the fact that every holomorphic mapping of $\mathbb{C}$ into it reduces to a constant. P.J. Kiernan has proved that $P_n\mathbb{C}$ minus $2n$ hyperplanes in general position is not hyperbolic, (see his paper "Hyperbolic submanifolds of complex projective space", to appear in Proc. Amer. Math. Soc.). This is why we required $\rho(n)$ to be at least $(2n+1)$ in the above. Motivated by Proposition 5.20, we conclude these notes with the following <u>conjecture</u>: $P_n\mathbb{C}$ minus $\rho(n)$ hyperplanes in general position is hyperbolic.

REFERENCES

[1] L.V. Ahlfors, The theory of meromorphic curves, Acta Soc. Sci. Fenn. Ser. A, vol. 3, No. 4.

[2] S.S. Chern, Complex manifolds without potential theory, Princeton, 1967.

[3] R.C. Gunning and R. Narasimhan, Immersions of open Riemann Surfaces, Math. Ann. 174(1967), 103-108.

[4] W.K. Hayman, Meromorphic functions, Oxford, 1964.

[5] W.E. Jenner, Rudiments of algebraic geometry, New York, 1963.

[6] R. Nevanlinna, Eindeutige analytische Funktionen, Berlin, 1953.

[7] H. Weyl and J. Weyl, Meromorphic functions and analytic curves, Princeton, 1943.

[8] H. Wu, Mappings of Riemann Surfaces (Nevanlinna theory), Proc. Sympos. Pure Math. vol. XI, "Entire Functions and Related Parts of Analysis." Amer. Math. Soc. 1968, 480-532.

INDEX OF PRINCIPAL DEFINITIONS

Ingram Content Group UK Ltd.
Milton Keynes UK
UKHW011242290523
422449UK00001B/89

9 780691 080734